AF560310

AN INTRODUCTION TO NANOTECHNOLOGY

NEW INDIA PUBLISHING AGENCY

New Delhi – 110 034

About the Authors

Dr. V. Ponnuswami, Ph. D. (Horticulture) is a presently the Dean of Horticultural College and Research Institute, Tamil Nadu AGricultural University, Periyakulam. He served as the visiting scientist in the Asian Vegetable Research and Development Centre (AVRDC), Taiwan during the year 1994-1995. He visited Bangkok, Hong Kong, Singapore and Philippines in connection with vegetable research. He has experience in teaching, research and extension work in Horticulture in various capacities for over 33 years. He has been honoured with four gold medals and five national level awards with JRF for his M.Sc. (Hort.) and Ph.D. (Hort.) programmes. He has released more than 21 crop varieties in Horticultural crops for benefit of farming community. He guided fifteen Ph. D.'s and fourteen M.Sc. student in the field of National and International journals, more than 200 popular articles and nineteen books. He is presently handling 18 externally mainly GOI funded schemes as Principal Investigator.

Dr. A. Rathinasamy, Ph.D. (Soil Science & Agrl. Chemistry) is presently working as Professor and Head, Department of Social Sciences, Horticultural College and Research Institute, Periyakulam. He has handled 40 courses for under graduate programme in different fields of soil science. He has guided 5 post graduate and 2 doctorate students of horticulture as advisory and National journals and 10 training manuals. He has trained the extension officials and farmers in INM and also under IAMWARM scheme.

Dr. C. Parameswari, Ph.D. is currently working as Assistant Professor (Plant Bredding and Genetics) in Horticultural College and Research Institute, Periyakulam. She is basically on oilseeds breeder with more than 10 years of post doctoral experience. She has research experience in conventional breeding and plant genetic engineering. She published 20 articles in National and International Journals. She has handled courses on nanotechnology and biotechnology for undergraduate students.

AN INTRODUCTION TO NANOTECHNOLOGY

A. Rathinasamy
Professor of Soil Science & Agricultural Chemistry
Horticultural College and Research Institute, Periyakulam

C. Parameswari
Assistant Professor (Plant Breeding & Genetics)
Horticultural College and Research Institute, Periyakulam

V. Ponnuswami
Dean
Horticultural College and Research Institute, Periyakulam

Tamil Nadu Agricultural University
Coimbatore - 641 003, Tamil Nadu

NEW INDIA PUBLISHING AGENCY
New Delhi – 110 034

NEW INDIA PUBLISHING AGENCY
101, Vikas Surya Plaza, CU Block, LSC Market
Pitam Pura, New Delhi – 110 034, India
Email: info@nipabooks.com
Web: www.nipabooks.com

For customer assistance, please contact
Phone: + 91-11-27 34 17 17 Fax: + 91-11- 27 34 16 16
E-Mail: feedbacks@nipabooks.com

ISBN : 978-93-81450-41-3

Composed and Designed by NIPA.

Digitally printed all across the globe.

Preface

The Study of manipulating materials on molecular as well as atomic scales is known as Nanotechnology. Nanotechnology deals with matters that are sized in between 1 to 100 nanometers. Nonotechnology is a new field of technology with science at nano levels. Materials behave very differently at nano levels. Their physical and chemical properties are unique at nano level. Today this field of advanced study has significant impact on all areas of the society and industry.

It can be said that Nanotechnology can be applied in almost all areas. The wide array of applications in nanotechnology has come to show one and all how much importance nanotechnology has in our lives today.

Ample apportunities are there to apply nanoscience and technology in the fields of physics, chemistry, agriculture, environmental sciences, fisheries, food industry, medicine, textiles, automobiles, etc. Because of these, many applications are available in Nanotechnology, the job opportunities are not at all scarce. This is one of the reasons why many people opt to study this couse.

Nanoscience and technology provides new ways and means to tackle critical issues and challenges in a very different manner for the benefit of mankind. This has to be understood by students, research scholars and scientists as well. The understanding of the nanomaterials and their properties are very essential for proper application in any field of science and technology.

Presently the scope for nanotechnology in India is quite huge. Numerous Indian firms use nanotechnology in the manufactures of their products. In addition to this, the Council of Scientific and Industrial Research (CSIR) has started as much as 28 laboratories just to increase the research possibilites of nanotechnology in India. Students of nanotechnology have the opportunity to get jobs are soon as they their education in this intersting field.

This book is written with an objective that students must understand the basic principles of nano science and technology for their further perspectives.

Authors

Contents

Chapter-1

What is Nanotechnology ?

What is Nanotechnology? Answers differ depending on who you ask, and their background. Broadly speaking however, nanotechnology is the act of purposefully manipulating matter at the atomic scale, otherwise known as the "nanoscale." Coined as "nano-technology" in a 1974 paper by Norio Taniguchi at the University of Tokyo, and encompassing a multitude of rapidly emerging technologies, based upon the scaling down of existing technologies to the next level of precision and miniaturization. Taniguchi approached nanotechnology from the 'top-down' standpoint, from the viewpoint of a precision engineer.

Foresight Nanotech Institute Founder K. Eric Drexler introduced the term "nanotechnology" to the world in 1986, using it to describe a 'bottom-up' approach. Drexler approaches nanotechnology from the point-of-view of a physicist, and defines the term as "large-scale mechano synthesis based on positional control of chemically reactive molecules." In the future, "nanotechnology" will likely include building machines and mechanisms with nanoscale dimensions, referred to these days as Molecular Nanotechnology (MNT).

It uses a basic unit of measure called a "nanometer" (abbreviated nm). Derived from the Greek word for midget, "nano" is a metric prefix and indicates a billionth part (10^{-9}).

There are one ***billion*** nms to a meter. Each nm is only three to five ***atoms*** wide. They're small, really small ~ 40,000 times smaller than the width of an average human hair.

Defining Nanotechnology

The term "nanotechnology" has been getting a lot of attention in the media in the last few years. News stories have heralded nanotechnology as the next scientific revolution—with promises of faster computers, cures for cancer and solutions to the energy crisis, to name a few. But what exactly is "nanotechnology"? And, can it fulfill all these promises?

The formal definition of nanotechnology from the National Nanotechnology Initiative (NNI) is:

Nanotechnology is the understanding and control of matter at dimensions between approximately 1 and 100 nanometers, where unique phenomena enable novel applications.

Encompassing nanoscale science, engineering, and technology, nanotechnology involves imaging, measuring, modeling, and manipulating matter at this length scale.

The NNI definition can be distilled to three basic concepts:

1. Nanotechnology is very, very small.

When something is on the nanoscale, it measures between 1 - 100 nanometers (nm) in at least one of its dimensions. When things are this small, they are much too small to see with our eyes, or even with a typical light microscope. Scientists have had to develop special tools, like scanning probe microscopes to see materials that are on the nanometer size scale.

Some materials have ALWAYS been on the nanoscale-like water molecules or silicon atoms. However, recently scientists have been able to use new tools and processes to synthesize and manipulate materials common at the macro scale to this size, like particles of TiO_2 (titanium dioxide).

2. At the nanoscale, materials may behave in different and unexpected ways.

At the nanoscale, many common materials exhibit unusual properties, such as remarkably lower resistance to electricity, lower melting points or faster chemical reactions.

For example, at the macro scale, gold (Au) is shiny and yellow. However, when the gold particles are 25 nm in size, they appear red. The smaller particles interact differently with light, so the gold particles appear a different color. Depending on the size and shape of the particles, gold can appear red, yellow or blue.

Fig. 1.1 : Gold coin

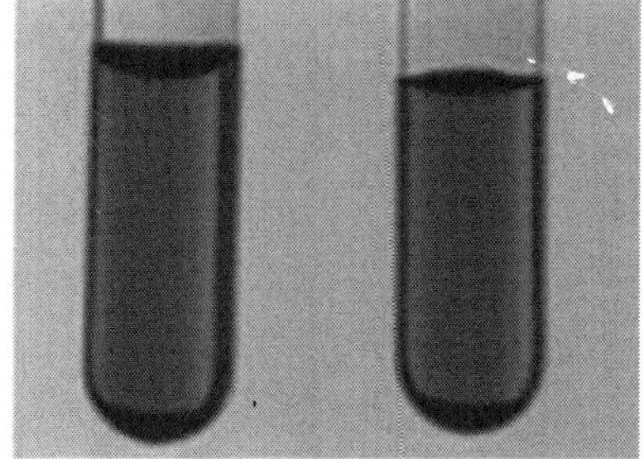

Fig. 1.2 : Nanoparticles of gold in solution

Another example of nanoparticles appearing different than the corresponding macro scale material is found in sunscreens. Titanium Dioxide (TiO_2) has been used in sun screens and sun blocks for a long time. It is one of the ingredients that make the creams appear white in color. Manufacturers are now using nanoparticles to create creams and gels that are clear-because nanoparticles of TiO_2 appear transparent.

Other properties can change when materials are on the nanoscale, too. For example, aluminum (Al) is the shiny pliable metal used to make soda cans. At the nanoscale, aluminum particles are extremely reactive and will explode. Nanoparticles are more reactive because they have more surface area than macro scale particles.

3. Researchers want to harness these different and unexpected behaviors to make new technologies.

By harnessing these new behaviors, researchers in many different disciplines hope to create many new things, ranging from everyday products, like antimicrobial socks and lighter tennis rackets to state of the art solar cells, faster, smaller computers or medical treatments that selectively treat the disease. Many scientists and engineers think that the possibilities are endless.

Is the iPod Nano an example of Nanotechnolgy?

Nanotechnology-enhanced products are already finding their way to the market. But, not every product labeled "nano" really utilizes nanotechnolgy. Some manufacturers are using the prefix "nano-" to communicate the relative small size of their specific products to potential buyers. In addition, they hope to capitalize the buzz and excitement that has surrounded nanotechnology.

For example, Indian carmakers recently introduced the Tata Nano:

Fig. 1.3 : image taken from http://en.wikipedia.org/wiki/Tata_Nano

This car is definitely not on the nanoscale—but the designers chose to name it "Nano" to communicate to potential buyers its use of "high technology" and its small size. This isn't the only example on the market today.

A more common example is the iPod Nano:

Fig. 1.4 : image taken from http://www.apple.com/ipodnano/

Again, this music player is not "nano-sized" but it is much smaller than the original iPod. However, it does utilize nanotechnology in the chips and circuits that make it work—the same way your laptop does!

There are some products out there that do use nanotechnology, even if it is not contained in the name! One of the most common instances is the use silver nanoparticles in consumer products. Silver is inherently anti-microbial and has been used to control bacteria since ancient times. By incorporating nanoscale silver into textiles, plastics, and household appliances, manufacturers can make materials that use a small amount of silver to kill bacteria without affecting other properties of the products.

Fig. 1.5 : "Benny the Bear"

Fig. 1.6 : plush toy Silver Seal computer keyboard

Fig. 1.7 : Sports Jersey from Sinotextiles Co,Ltd

(images taken from http://www.nanotechproject.org/inventories/consumer/)

However, there is concern that increased use of silver in this manner may pose an environmental risk or lead to the development of silver-antibiotic-resistant bacteria.

The Project on Emerging Nanotechnologies has created an inventory of consumer products that they believe actually utilize nanotechnology in some way. However, due to the proprietary nature of these products, a definitive decision on *exactly* how the product is utilizing nanotechnology can often not be made.

Can nanotechnology live up to the hype?

Nanotechnology has been dubbed "the next big thing". It is being heralded as the key to new cancer treatments, energy independence, improved electronics and bringing clean water to third world countries. With such a diverse range of possible applications, nanotechnology does have the capacity to change the world we live in, in the same way that computers have changed society over the last 30 years. But to accomplish these feats, scientists and engineers that work on nanoscale materials still have a lot of research to do—to better understand how nanoscale materials behave and how to synthesize them reliably. Some applications are closer to market than others. For instance, researchers, like Prof. Naomi Halas and Jennifer West at Rice University have collaborated to develop a cancer treatment using gold nano shells. The treatment is still under development, but may find its way to clinical trials in the near future.

History of Nanotechnology

Overview

In 1965, Gordon Moore, one of the founders of Intel Corporation, made the astounding prediction that the number of transistors that could be fit in a given area would double every 18 months for the next ten years. This it did and the phenomenon became known as Moore's Law. This trend has continued far past the predicted 10 years until this day, going from just over 2000 transistors in the original 4004 processors of 1971 to over 700,000,000 transistors in the Core 2. There has, of course, been a corresponding decrease in the size of individual electronic elements, going from millimeters in the 60s to hundreds of nanometers in modern circuitry.

At the same time, the chemistry, biochemistry and molecular genetics communities have been moving in the other direction. Over much the same period, it has become possible to direct the synthesis, either in the test tube or in modified living organisms,

Now, at the beginning of a new century, three powerful technologies have met on a common scale — the nanoscale — with the promise of revolutionizing both the worlds of electronics and of biology. This new field, which we refer to as biomolecular nanotechnology, holds many possibilities from fundamental research in molecular biology and biophysics to applications in biosensing, biocontrol, bioinformatics, genomics, medicine, computing, information storage and energy conversion.

Historical background

Humans have unwittingly employed nanotechnology for thousands of years, for example in making steel, paintings and in vulcanizing rubber. Each of these processes rely on the properties of stochastically-formed atomic ensembles mere nanometers in size, and are distinguished from chemistry in that they don't rely on the properties of individual molecules. But the development of the body of concepts now subsumed under the term nanotechnology has been slower.

The first mention of some of the distinguishing concepts in nanotechnology (but predating use of that name) was in 1867 by James Clerk Maxwell when he proposed as a thought experiment a tiny entity known as Maxwell's Demon able to handle individual molecules.

The first observations and size measurements of nano-particles was made during the first decade of the 20th century. They are mostly associated with Richard Adolf Zsigmondy who made a detailed study of gold sols and other nanomaterials with sizes down to 10 nm and less. He published a book in 1914. He used ultramicroscope that employes the *dark field* method for seeing particles with sizes much less than light wavelength. Zsigmondy was also the first who used **nanometer** explicitly for characterizing particle size. He determined it as 1/1,000,000 of millimeter. He developed the first system classification based on particle size in the nanometer range.

There have been many significant developments during the 20th century in characterizing nanomaterials and related phenomena, belonging to the field of interface and colloid science. In the 1920s, Irving Langmuir and Katharine B. Blodgett introduced the concept of a monolayer, a layer of material one molecule thick. Langmuir won a Nobel Prize in chemistry for his work. In the early 1950s, Derjaguin and Abrikosova conducted the first measurement of surface forces.

There have been many studies of *periodic colloidal structures* and principles of molecular self-assembly that are overviewed in the following chapters. There are many other discoveries that serve as the scientific basis for the modern nanotechnology which can be found in the "Fundamentals of Interface and Colloid Science by H.Lyklema.

Conceptual origins

The topic of nanotechnology was again touched upon by "There's Plenty of Room at the Bottom," a talk given by physicist Richard Feynman at an American Physical Society meeting at Caltech on December 29, 1959. Feynman described a process by which the ability to manipulate individual atoms and molecules might be developed, using one set of precise tools to build and operate another proportionally smaller set, so on down to the needed scale. In the course of this, he noted, scaling issues would arise from the changing magnitude of various physical phenomena: gravity would become less important, surface tension and Van der Waals attraction would become more important, etc. This basic idea appears feasible, and exponential assembly enhances it with parallelism to produce a useful quantity of end products.

In 1965 Gordon Moore observed that silicon transistors were undergoing a continual process of scaling downward, an observation which was later codified as Moore's law. Since his observation transistor minimum feature sizes have decreased from 10 micrometers to the 45-65 nm range in 2007; one minimum feature is thus roughly 180 silicon atoms long.

The term "nanotechnology" was first defined by Norio Taniguchi of the Tokyo Science University in a 1974 paper as follows: "'Nano-technology' mainly consists of the processing of, separation, consolidation, and deformation of materials by one atom or one molecule." Since that time the definition of nanotechnology has generally been extended to include features as large as 100 nm. Additionally, the idea that nanotechnology embraces structures exhibiting quantum

60 nm

As soon as I mention this, people tell me about miniaturization, and how far it was progressed today. They tell me about, electric motors that are the size of the nail of your small finger. And there is a device on the market, they tell me, by which you can write the Lord's Prayer on the head of pin. But, that's the most primitive, halting step in the direction I intend to discuss. It is staggeringly small world that is below. In the year 2000, when they look back at this age, they will wonder why it was not until the year 1950 that anybody began seriously to move in this direction.

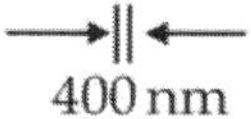

Richard P. Feynman, 1960

From "There's Plenty of Room at the Bottom" December 29th, 1959

mechanical aspects, such as quantum dots, has further evolved its definition.

Also in 1974 the process of atomic layer deposition, for depositing uniform thin films one atomic layer at a time, was developed and patented by Dr. Tuomo Suntola and co-workers in Finland.

In the 1980s, handling of individual atoms and molecules was conceptually explored in depth by Dr. K. Eric Drexler, who promoted the technological significance of nano-scale phenomena and devices through speeches and books. Drexler's vision of nanotechnology is often called "Molecular Nanotechnology" (MNT) or "molecular manufacturing."

In 2004, Richard Jones wrote a book called Soft Machines (nanotechnology and life), which is a book about nanotechnology for the general reader, published by Oxford University. In this book he describes radical nanotechnology as a deterministic/mechanistic idea of nano engineered machines that does not take into account the nanoscale challenges such as wetness, stickness, brownian motion, high viscosity (Drexler view). He also explains what is soft nanotechnology or more appropriatelly biomimetic nanotechnology which is the way forward, if not the best, to design functional nano devices that can cope with all the problems at nanoscale.

Experimental advances

Nanotechnology and nanoscience got a boost in the early 1980s with two major developments: the birth of cluster science and the invention of the scanning tunneling microscope (STM). This development led to the discovery of fullerenes in 1985 and the structural assignment of carbon nanotubes a few years later. In another development, the synthesis and properties of semiconductor nanocrystals were studied. This led to a fast increasing number of semiconductor nanoparticles of quantum dots.

In the early 1990s Huffman and Kraetschmer, of the University of Arizona, discovered how to synthesize and purify large quantities of fullerenes. This opened the door to their characterization and functionalization by hundreds of investigators in government and industrial laboratories. Shortly after, rubidium doped C_{60} was found to be a mid temperature (Tc = 32 K) superconductor. At a meeting of the Materials Research Society in 1992, Dr. T. Ebbesen described to a spellbound audience his discovery and characterization of carbon nanotubes. This event sent those in attendance and others downwind of his presentation into their laboratories to reproduce and push those discoveries forward. Using the same or similar tools as those used by Huffman and Kratschmere, hundreds of researchers further developed the field of nanotube-based nanotechnology.

At present the practice of nanotechnology embraces both stochastic approaches (in which, for example, supramolecular chemistry creates waterproof pants) and deterministic approaches wherein single molecules (created by stochastic chemistry) are manipulated on substrate surfaces (created by stochastic deposition methods) by deterministic methods comprising nudging them with STM or AFM probes and causing simple binding or cleavage reactions to occur. The dream of a complex, deterministic molecular nanotechnology remains elusive. Since the mid 1990s, thousands of surface scientists and thin film technocrats have latched on to the nanotechnology bandwagon and redefined their disciplines as nanotechnology. This has caused much confusion in the field and has spawned thousands of "nano"-papers on the peer reviewed literature. Most of these reports are extensions of the more ordinary research done in the parent fields.

For the future, some means has to be found for MNT design evolution at the nanoscale which mimics the process of biological evolution at the molecular scale. Biological evolution proceeds by random variation in ensemble averages of organisms combined with culling of the less-successful variants and reproduction of the more-successful variants, and macro scale engineering design also proceeds by a process of design evolution from simplicity to complexity as set forth somewhat satirically by John Gall: "A complex system that works is invariably found to have evolved from a simple system that worked. A complex system designed from scratch never works and can not be patched up to make it work. You have to start over, beginning with a system that works." A breakthrough in MNT is needed which proceeds from the simple atomic ensembles which can be built with, e.g., an STM to complex MNT systems via a process of design evolution. A handicap in this process is the difficulty of seeing and manipulation at the nanoscale compared to the macro scale which makes deterministic selection of successful trials difficult; in contrast biological evolution proceeds via action of what Richard Dawkins has called the "blind watchmaker" comprising random molecular variation and deterministic reproduction/extinction.

Introduction to Nanotechnology

Nanotechnology is defined as the study and use of structures between 1 nanometer and 100 nanometers in size. To give you an idea of how small that is, it would take eight hundred 100 nanometer particles side by side to match the width of a human hair.

Scientists have been studying and working with nanoparticles for centuries, but the effectiveness of their work has been hampered by their inability to see the structure of nanoparticles. In recent decades the development of microscopes capable of displaying particles as small as atoms has allowed scientists to see what they are working with.

The following illustration titled "The Scale of Things", created by the U. S. Department of Energy, provides a comparison of various objects to help you begin to envision exactly how small a nanometer is. The chart starts with objects that can be seen by the unaided eye, such as an ant, at the top of the chart, and progresses to objects about a nanometer or less in size, such as the ATP molecule used in humans to store energy from food.

Now that you have an idea of how small a scale nanotechnologists work with, consider the challenge they face. Think about how difficult it is for many of us to insert thread through the eye of a needle. Such an image helps you imagine the problem scientists have working with nanoparticles that can be as much as one millionth the size of the thread. Only through the use of powerful microscopes can they hope to 'see' and manipulate these nano-sized particles.

The ability to see nano-sized materials has opened up a world of possibilities in a variety of industries and scientific endeavors. Because nanotechnology is essentially a set of techniques that allow manipulation of properties at a very small scale, it can have many applications, which are dealt with later.

It is useful to know the approximate sizes of things. Is something bigger than a bread box? Will it fit through a doorway? Is it as big as it is supposed to be? These are all questions that we may find ourselves asking on a regular basis. The two systems of measurement used today are the English and Metric System.

Fig. 1.8 : The Scale of Things – Nanometers and More

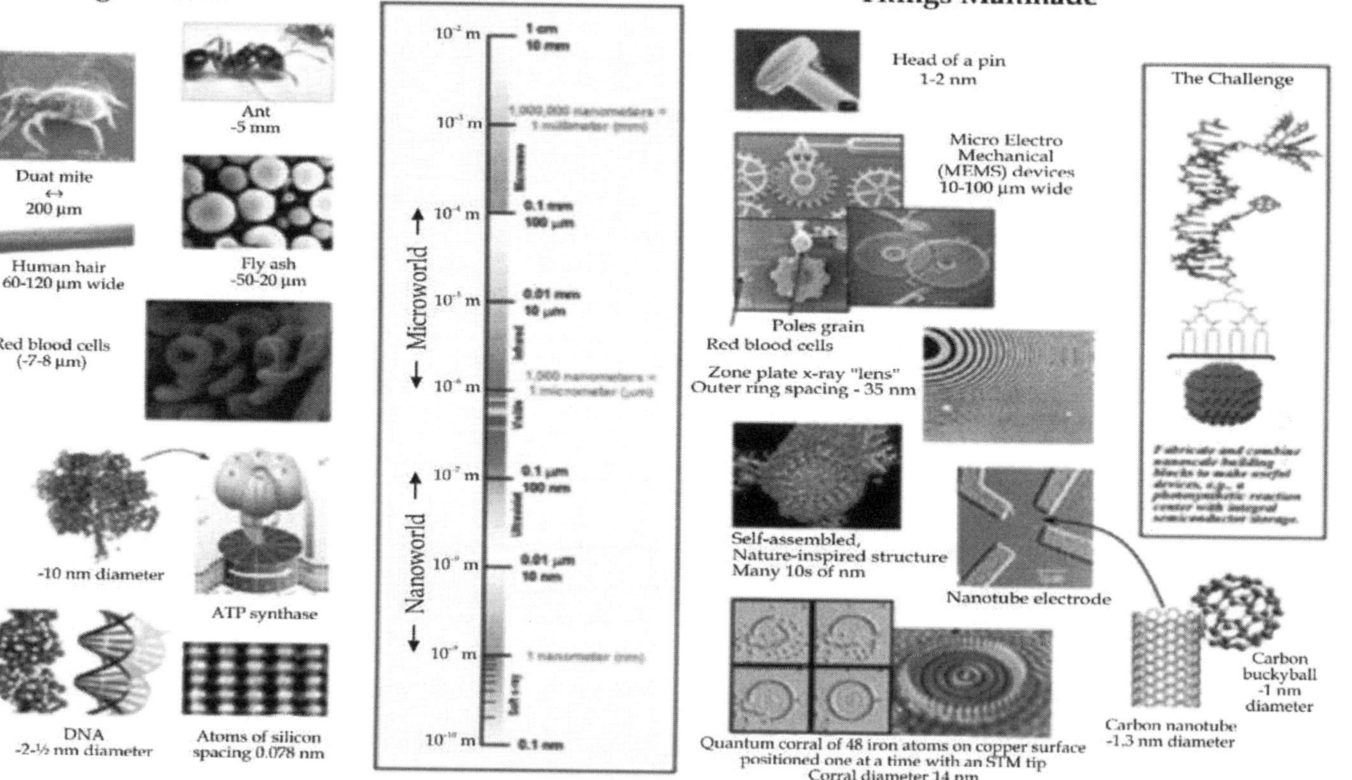

In everyday life the English system of measurement with units like feet, gallons and pounds is primarily used in the United States. Scientists and the rest of the world use the metric system of measurement with base units such as the meter, liter or gram. The base unit is then multiplied or divided by powers of ten to represent larger or smaller measurements. Prefixes are used to relate the relative size of the measurement with respect to the base unit. Since the base unit for length is the meter the table below shows all the different prefixes that could be used to express length.

Prefix	Measurement	Scientific Notation
Kilo-	1000 m	1×10^{3} m
Hecta-	100 m	1×10^{2} m
Deka-	10 m	1×10^{1} m
BASE	1 m	1×10^{0} m
Deci-	0.1m	1×10^{-1}m
Centi-	0.01 m	1×10^{-2} m
Milli-	0.001m	1×10^{-3} m
Micro-	0.000001 m	1×10^{-6} m
Nano-	0.000000001 m	1×10^{-9} m
Pico-	0.000000000001 m	1×10^{-12} m
Femto-	0.000000000000001 m	1×10^{-15} m

1000 meters would be written as 1 kilometer. Other base units can be used as well. For mass, the base unit is the gram, so 1/100th of a gram would be written as a 1 milligram. Often is it hard to visualize what those numbers really mean-especially since base units in the metric systems are not something we may use in our every day life.

How big is a kilometer?

The world's longest suspension bridge, Akashi Kaikyo Bridge, is just less than 2 kilometer (km) long, measuring 1,991 m. It would take you approximately 23 minutes to walk across this bridge. The Sears Tower, the tallest building in North America, measures ~0.5 km (527 m) to the top of the antennae.

Fig. 1.9 : image taken from http://en.wikipedia.org/wiki/Akashi-Kaikyo_Bridge

Fig. 1.10 : images taken from http://en.wikipedia.org/wiki/Sears_tower

How big is a meter?

A meter is a little bit bigger than a yard. A lot of things in our everyday lives are around the same size as a meter. An elementary school student is one meter tall. However, most professional basketball players are at least 2 meters tall. Shaquille O'Neill, at 7'1", is 2.1 meters tall. He seems to tower over President Bush.

Fig.1.11 : image taken from http://www.themomsbuzz.com/moms_buzz/children/index.html

Fig. 1.12 : images taken from http://en.wikipedia.org/wiki/Sears_tower

How big a centimeter?

A centimeter (cm) is 1/100th of a meter-meaning that if you cut up a meter stick into 100 equal pieces, each piece would be one centimeter in size.

Your pinky finger is about 1 cm wide. Same goes for the width of the sugar cube.

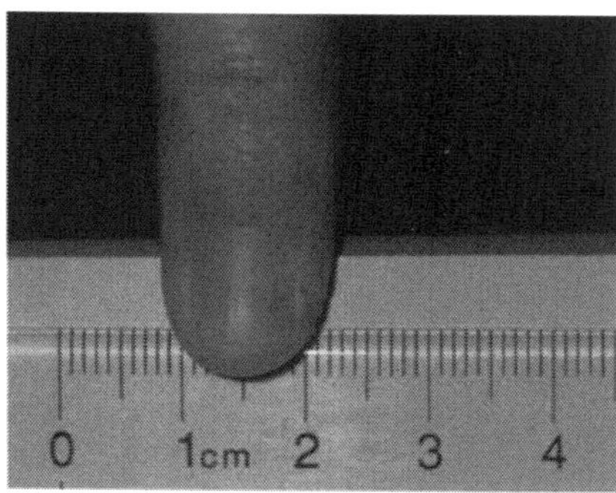

Fig. 1. 13 : Pinky finger

Fig. 1.14 : Sugar cubes

How big is a millimeter?

A millimeter (mm) is 1/1000th of a meter. A dime is approximately 1 mm thick. Grains of sand range from 0.1 mm to 2 mm in size.

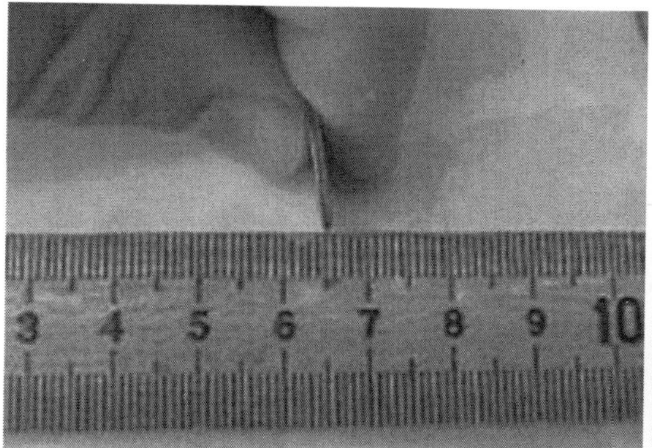

Fig. 1.15 : Fiimage taken from http://www.lcurve.org/millbill.htm

Fig. 1.16 : images taken from http://en.wikipedia.org/wiki/Sand

How big is a micrometer?

A micrometer, also called a micron, is one thousand times smaller than millimeter. It is equal to 1/1,000,000th (or one millionth of meter). Things on this scale usually can't be seen with your eyes.

Fig. 1.17 : Human hair magnified 200X

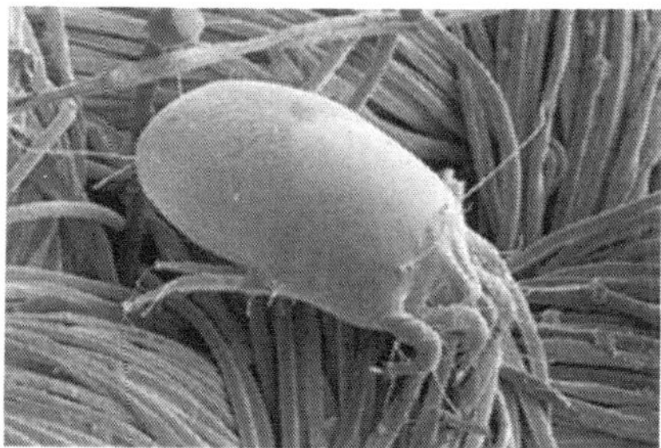

Fig. 1.18 : Dust mite magnified 40X

The diameter of a hair, which is 40-50 microns wide, is very hard to discern without the use of a magnifying glass. A magnifying glass will help you see a dust mite. Dust mites are usually around 400 microns long.

Things that are just a few microns in size can't be seen with a magnifying glass. However, a light microscope, like the ones you may have used in biology class, can help you see things this small. Red Blood cells are 6-10 microns in diameter and WBCs are 10 microns in diameter. Many types of bacteria typically measure 5-20 microns. White blood cells are ~10,000 nm in diameter

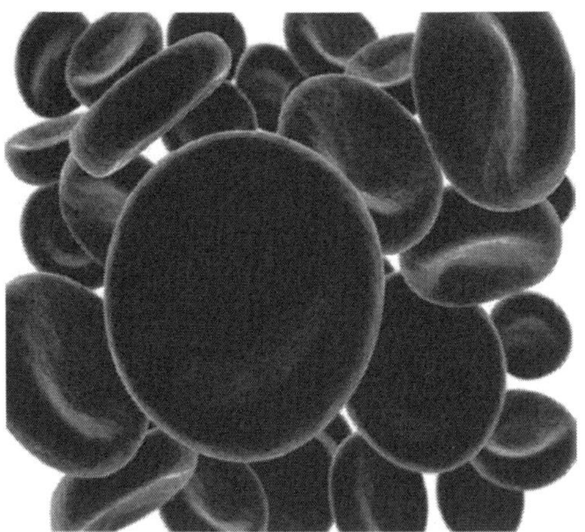

Fig. 1.19 : Red blood cells magnified 1700X

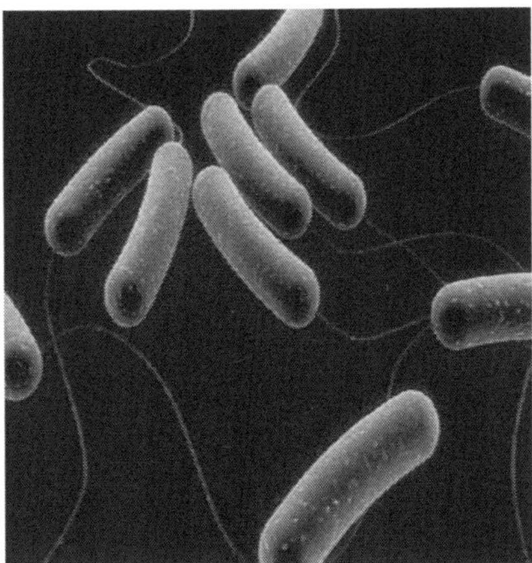

Fig. 1.20 : Bacteria magnified ##X

How big is a nanometer?

A nanometer (nm) is 1,000 times smaller than a micrometer. It is equal to 1/1,000,000,000th or one-billionth of a meter. When things are this small, you can't see them with your eyes, or a light microscope. Objects this small require a special tool called a scanning probe microscope.

All these things are on the nanometer scale: Virus (30-50 nm), DNA (2.5 nm), buckyballs (~1 nm in diameter), CNT (~1 nm in diameter), hydrogen atom (0 .1 nm), Quantum Dots of CdSe (8 nm), Dendrimers (~10 nm), proteins (5 to 50 nm).

Atoms are smaller than a nanometer. One atom measures ~0.1-0.3 nm, depending on the element.

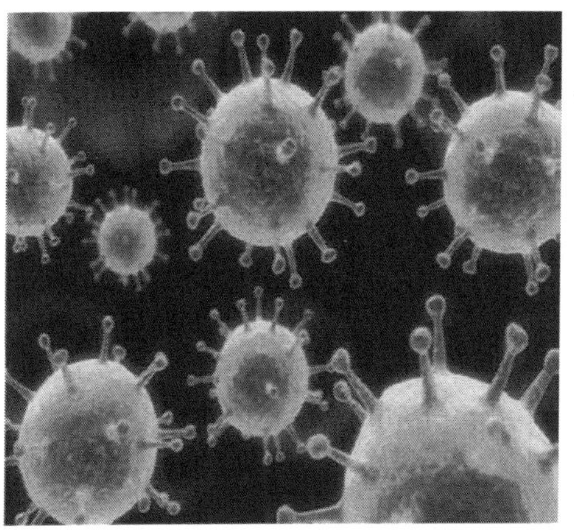

Fig. 1.21 : Virus magnified 20,000X

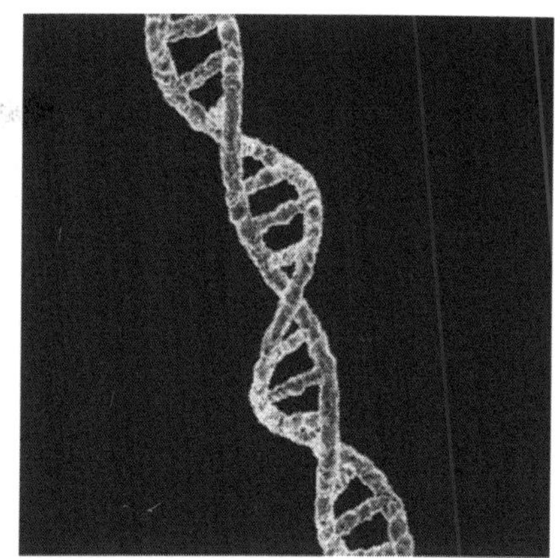

Fig. 1.22 : Model of DNA

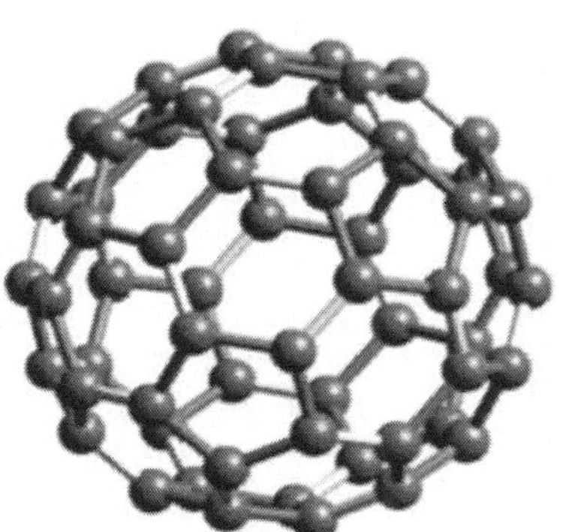

Fig. 1.23 : Model of a buckyball

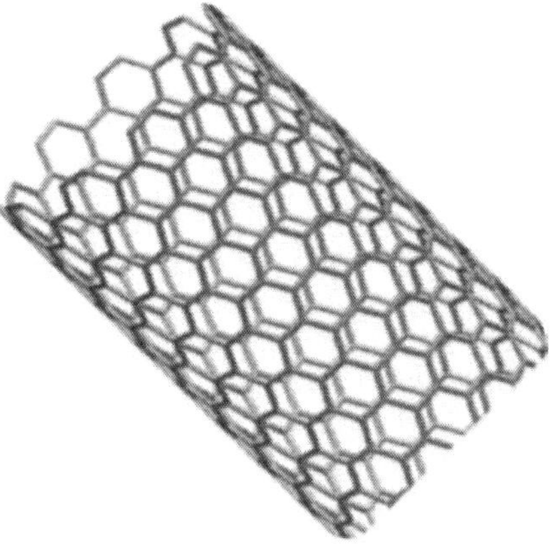

Fig. 1.24 : model of a carbon nanotube

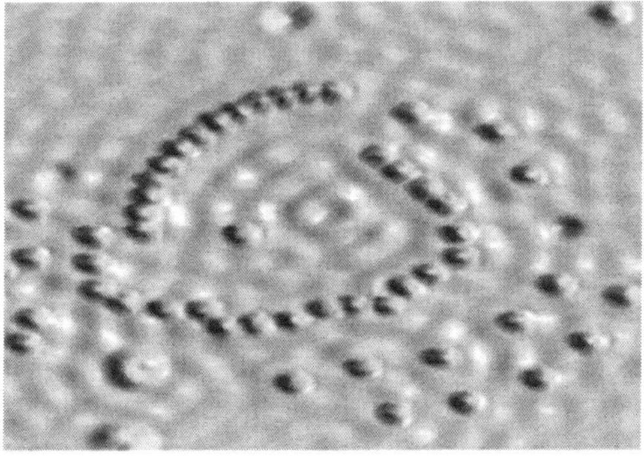

Fig. 1.25 : SEM image of atoms on a surface

Here are some other everyday objects measured in nanometers:

- One inch equals 25.4 million nanometers.
- A sheet of paper is about 100,000 nanometers thick.
- A human hair measures roughly 50,000 to 100,000 nanometers in diameter.
- Your fingernails grow one nanometer every second.

Importance of Moore's Law

Moore's Law describes a long-term trend in the history of computing hardware, in which the number of transistors that can be placed inexpensively on an integrated circuit has doubled approximately every two years. Rather than being a naturally-occurring "law" that cannot be controlled, however, Moore's Law is effectively a business practice in which the advancement of transistor counts occurs at a fixed rate.

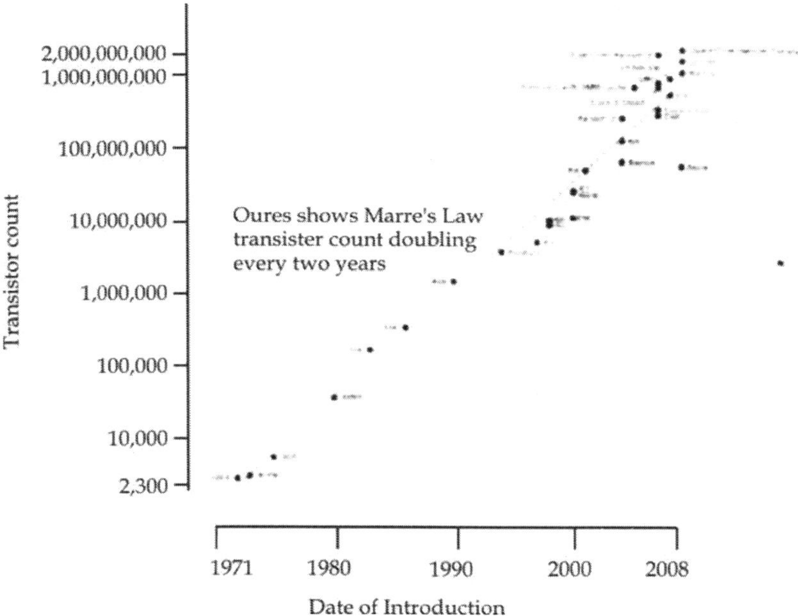

Fig. 1.26 : CPU Transistor Counts 1971-2008 & Moore's Law

The capabilities of many digital electronic devices are strongly linked to Moore's law: processing speed, memory capacity, sensors and even the number and size of pixels in digital cameras. All of these are improving at (roughly) exponential rates as well. This has dramatically increased the usefulness of digital electronics in nearly every segment of the world economy. Moore's law precisely describes a driving force of technological and social change in the late 20th and early 21st centuries. The trend has continued for more than half a century and is not expected to stop until 2015 or even later.

The law is named for Intel co-founder Gordon E. Moore, who introduced it in a 1965 paper. It has since been used in the semiconductor industry to guide long term planning and to set targets for research and development.

Transistors per integrated circuit

The most popular formulation is of the doubling of the number of transistors on integrated circuits every two years. At the end of the 1970s, Moore's law became the limit for the number of transistors on the most complex chips. Recent trends show that this rate has been maintained into 2007.

Density at minimum cost per transistor

This is the formulation given in Moore's 1965 paper. It is not just about the density of transistors that can be achieved, but about the density of transistors at which the cost per transistor is the lowest. As more transistors are put on a chip, the cost to make each transistor decreases, but the chance that the chip will not work due to a defect increases. In 1965, Moore examined the density of transistors at which cost is minimized, and observed that, as transistors were made smaller through advances in photolithography, this number would increase at "a rate of roughly a factor of two per year".

Future trends

Computer industry technology "road maps" predict (as of 2001[update]) that Moore's law will continue for several chip generations. Depending on and after the doubling time used in the calculations, this could mean up to a hundredfold increase in transistor count per chip within a decade. The semiconductor industry technology roadmap uses a three-year doubling time for microprocessors, leading to a tenfold increase in the next decade. Intel was reported in 2005 as stating that the downsizing of silicon chips with good economics can continue during the next decade and in 2008 as predicting the trend through 2029.

Ultimate limits of the law

On 13 April 2005, Gordon Moore stated in an interview that the law cannot be sustained indefinitely: "It can't continue forever. The nature of exponentials is that you push them out and eventually disaster happens." He also noted that transistors would eventually reach the limits of miniaturization at atomic levels:

In terms of size [of transistors] you can see that we're approaching the size of atoms which is a fundamental barrier, but it'll be two or three generations before we get that far—but that's as far out as we've ever been able to see. We have another 10 to 20 years before we reach a fundamental limit. By then they'll be able to make bigger chips and have transistor budgets in the billions.

In January 1995, the Digital Alpha 21164 microprocessor had 9.3 million transistors. This 64-bit processor was a technological spearhead at the time, even if the circuit's market share remained average. Six years later, a state of the art microprocessor contained more than 40 million transistors. It is theorized that with further miniaturization, by 2015 these processors should contain more than 15 billion transistors, and by 2020 will be in molecular scale production, where each molecule can be individually positioned.

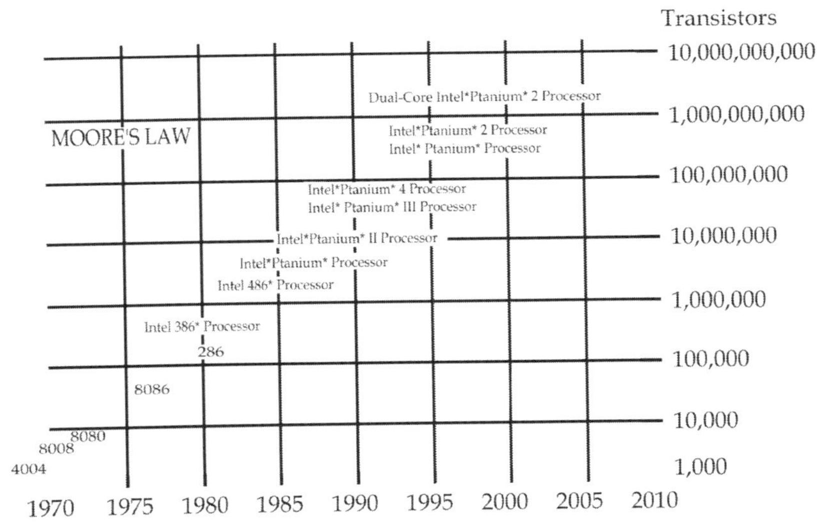

Fig. 1.27 : Moore's law

In 2003 Intel predicted the end would come between 2013 and 2018 with 16 nanometer manufacturing processes and 5 nanometer gates, due to quantum tunneling, although others suggested chips could just get bigger, or become layered. In 2008 it was noted that for the last 30 years it has been predicted that Moore's law would last at least another decade.

Some see the limits of the law as being far in the distant future. Lawrence Krauss and Glenn D. Starkman announced an ultimate limit of around 600 years in their paper, based on rigorous estimation of total information-processing capacity of any system in the Universe.

Then again, the law has often met obstacles that appeared insurmountable and, before long, surmounted them. In that sense, Moore says he now sees his law as more beautiful than he had realized: "Moore's law is a violation of Murphy's law. Everything gets better and better."

❑❑❑

Chapter-2

Nanomaterials and its Applications

Nanomaterials are a field which takes a materials science-based approach to nanotechnology. It studies materials with morphological features on the nanoscale, and especially those which have special properties stemming from their nanoscale dimensions. Nanoscale is usually defined as smaller than a one tenth of a micrometer in at least one dimension, though this term is sometimes also used for materials smaller than one micrometer.

An aspect of nanotechnology is the vastly increased ratio of surface area to volume present in many nanoscale materials which makes possible new quantum mechanical effects, for example the "quantum size effect" where the electronic properties of solids are altered with great reductions in particle size. This effect does not come into play by going from macro to micro dimensions. However, it becomes pronounced when the nanometer size range is reached. A certain number of physical properties also alter with the change from macroscopic systems. Novel mechanical properties of nanomaterials are a subject of nanomechanics research. Catalytic activities also reveal new behaviour in the interaction with biomaterials.

As mentioned above, materials reduced to the nanoscale can suddenly show very different properties compared to what they exhibit on a macro scale, enabling unique applications. For instance, opaque substances become transparent (copper); inert materials attain catalytic properties (platinum); stable materials turn combustible (aluminum); solids turn into liquids at room temperature (gold); insulators become conductors (silicon). Materials such as gold, which is chemically inert at normal scales, can serve as a potent chemical catalyst at nanoscales. Much of the fascination with nanotechnology stems from these unique quantum and surface phenomena that matter exhibits at the nanoscale.

Tools and techniques

The first observations and size measurements of nano-particles were made during the first decade of the 20th century. They are mostly associated with the name of Zsigmondy who made detailed studies of gold sols and other nanomaterials with sizes down to 10 nm and less. He published a book in 1914. He used ultramicroscope that employs a *dark field* method for seeing particles with sizes much less than light wavelength.

There are traditional techniques developed during 20th century in Interface and Colloid Science for characterizing nanomaterials. These are widely used for *first generation* passive nanomaterials specified in the next section.

These methods include several different techniques for characterizing particle size distribution. This characterization is imperative because many materials that are expected to be nano-sized are actually aggregated in solutions. Some of methods are based on light scattering. Other applies ultrasound, such as ultrasound attenuation spectroscopy for testing concentrated nano-dispersions and micro emulsions.

There is also a group of traditional techniques for characterizing surface charge or zeta potential of nano-particles in solutions. This information is required for proper system stabilization, preventing its aggregation or flocculation. These methods include microelectrophoresis, electrophoretic light scattering and electroacoustics.

Materials used in nanotechnology

Materials referred to as "nanomaterials" generally fall into two categories: fullerenes, and inorganic nanoparticles.

Fullerenes

Buckminsterfullerene C_{60}, also known as the buckyball, is the smallest member of the fullerene family. The fullerenes are a class of allotropes of carbon which conceptually are graphene sheets rolled into tubes or spheres. These include the carbon nanotubes which are of interest both because of their mechanical strength and also because of their electrical properties. For the past decade, the chemical and physical properties of fullerenes have been a hot topic in the field of research and development, and are likely to continue to be for a long time. In April 2003, fullerenes were under study for potential medicinal use: binding specific antibiotics to the structure of resistant bacteria and even target certain types of cancer cells such as melanoma. The October 2005 issue of Chemistry and Biology contains an article describing the use of fullerenes as light-activated antimicrobial agents. In the field of nanotechnology, heat resistance and superconductivity are among the properties attracting intense research.

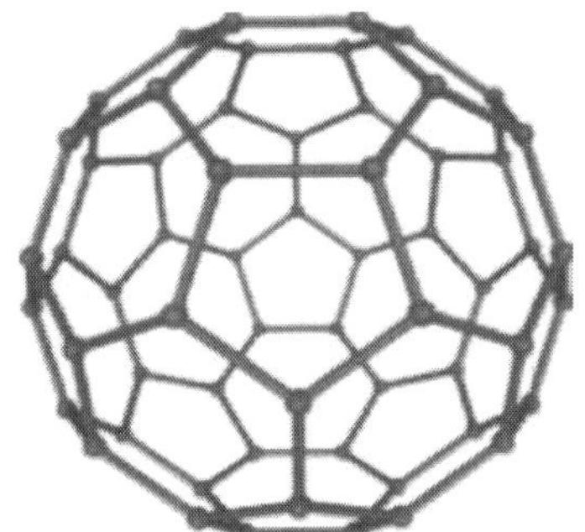
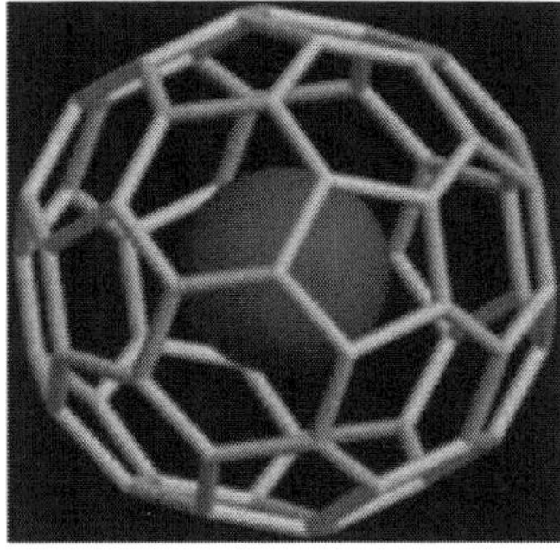

Fig. 2.1 : Fullereaes

A common method used to produce fullerenes is to send a large current between two nearby graphite electrodes in an inert atmosphere. The resulting carbon plasma arc between the electrodes cools into sooty residue from which many fullerenes can be isolated.

Nanoparticles

Nanoparticles or nanocrystals made of metals, semiconductors, or oxides are of particular interest for their mechanical, electrical, magnetic, optical, chemical and other properties. Nanoparticles have been used as quantum dots and as chemical catalysts. Nanoparticles are of great scientific interest as they are effectively a bridge between bulk materials and atomic or molecular structures. A bulk material should have constant physical properties regardless of its size, but at the nano-scale this is often not the case. Size-dependent properties are observed such as quantum confinement in semiconductor particles, surface plasmon resonance in some metal particles and super paramagnetism in magnetic materials.

Nanoparticles exhibit a number of special properties relative to bulk material. For example, the bending of bulk copper (wire, ribbon, etc.) occurs with movement of copper atoms/clusters at about the 50 nm scale. Copper nanoparticles smaller than 50 nm are considered super hard materials that do not exhibit the same malleability and ductility as bulk copper. The change in properties is not always desirable. Ferroelectric materials smaller than 10 nm can switch their magnetization direction using room temperature thermal energy, thus making them useless for memory storage. Suspensions of nanoparticles are possible because the interaction of the particle surface with the solvent is strong enough to overcome differences in density, which usually result in a material either sinking or floating in a liquid. Nanoparticles often have unexpected visual properties because they are small enough to confine their electrons and produce quantum effects. For example gold nanoparticles appear deep red to black in solution.

The often very high surface area to volume ratio of nanoparticles provides a tremendous driving force for diffusion, especially at elevated temperatures. Sintering is possible at lower temperatures and over shorter durations than for larger particles. This theoretically does not affect the density of the final product, though flow difficulties and the tendency of nanoparticles to agglomerate do complicate matters. The surface effects of nanoparticles also reduce the incipient melting temperature.

Nanocrystal

Fahlman, B. D. has described a **nanocrystal** as any nanomaterial with at least one dimension d" 100nm and that is single crystalline. More properly, any material with a dimension of less than 1 micrometre, i.e., 1000 nanometers, should be referred to as a nanoparticle, not a nanocrystal. For example, any particle which exhibits regions of crystallinity should be termed nanoparticle or nanocluster based on dimensions. These materials are of huge technological interest since many of their electrical and thermodynamic properties show strong size dependence and can therefore be controlled through careful manufacturing processes.

Crystalline nanoparticles are also of interest because they often provide single-domain crystalline systems that can be studied to provide information that can help explain the behaviour of macroscopic samples of similar materials, without the complicating presence of grain boundaries and other defects. Semiconductor nanocrystals in the sub-10nm size range are often referred to as quantum dots.

Crystalline nanoparticles made with zeolite are used as a filter to turn crude oil onto diesel fuel at an ExxonMobil oil refinery in Louisiana, a method cheaper than the conventional way. A layer of crystalline nanoparticles is used in a new type of solar panel named SolarPly made by Nanosolar. It is cheaper than other solar panels, more flexible, and claims 12% efficiency. (Conventionally inexpensive organic solar panels convert 9% of the sun's energy into electricity). Crystal tetrapods 40 nanometers wide convert photons into electricity, but only have 3% efficiency. (Source: National Geographic June 2006)

Better electroluminiscence has been observed in Silicon nanocrystals. Rare earth incorporation in Si nanocrystal has been reported to produce a satisfactory optical gain. The term NanoCrystal is a registered trademark of Elan Pharma International Limited (Ireland) used in relation to Elan's proprietary milling process and nanoparticulate drug formulations.

Nanocomposite

A **nanocomposite** can be defined as a multiphase solid material where one of the phases has one, two or three dimensions of less than 100 nanometers (nm), or structures having nano-scale repeat distances between the different phases that make up the material. In the broadest sense this definition can include porous media, colloids, gels and copolymers, but is more usually taken to mean the solid combination of a bulk matrix and nano-dimensional phase(s) differing in properties due to dissimilarities in structure and chemistry. The mechanical, electrical, thermal, optical, electrochemical, catalytic properties of the nanocomposite will differ markedly from that of the component materials. Size limits for these effects have been proposed, <5 nm for catalytic activity, <20 nm for making a hard magnetic material soft, <50 nm for refractive index changes, and <100 nm for achieving superparamagnetism, mechanical strengthening or restricting matrix dislocation movement.

Nanocomposites are found in nature, for example in the structure of the abalone shell and bone. The use of nanoparticle-rich materials long predates the understanding of the physical and chemical nature of these materials.

In mechanical terms, nanocomposites differ from conventional composite materials due to the exceptionally high surface to volume ratio of the reinforcing phase and/or its exceptionally high aspect ratio. The reinforcing material can be made up of particles (e.g. minerals), sheets (e.g. exfoliated clay stacks) or fibres (e.g. carbon nanotubes or electrospun fibres). The area of the interface between the matrix and reinforcement phase(s) is typically an order of magnitude greater than for conventional composite materials. The matrix material properties are significantly affected in the vicinity of the reinforcement.

This large amount of reinforcement surface area means that a relatively small amount of nanoscale reinforcement can have an observable effect on the macroscale properties of the composite. For example, adding carbon nanotubes improves the electrical and thermal conductivity. Other kinds of nanoparticulates may result in enhanced optical properties, dielectric properties, heat resistance or mechanical properties such as stiffness, strength and resistance to wear and damage. In general, the nano reinforcement is dispersed into the matrix during processing. The percentage by weight (called *mass fraction*) of the nanoparticulates introduced can remain very low (on the order of 0.5% to 5%) due to the low filler percolation threshold, especially for the most commonly used non-spherical, high aspect ratio fillers (e.g. nanometer-thin platelets, such as clays, or nanometer-diameter cylinders, such as carbon nanotubes).

Ceramic-matrix nanocomposites

In this group of composites the main part of the volume is occupied by a ceramic, i.e. a chemical compound from the group of oxides, nitrides, borides, silicides etc. In most cases, ceramic-matrix nanocomposites encompass a metal as the second component. Ideally both components, the metallic one and the ceramic one, are finely dispersed in each other in order to elicit the particular nanoscopic properties. Nanocomposite from these combinations was demonstrated in improving their optical, electrical and magnetic properties as well as tri biological, corrosion-resistance and other protective properties.

The binary phase diagram of the mixture should be considered in designing ceramic-metal nanocomposites and measures have to be taken to avoid a chemical reaction between both components. The last point mainly is of importance for the metallic component that may easily react with the ceramic and thereby loosing its metallic character. This is not an easily obeyed constraint, because the preparation of the ceramic component generally requires high process temperatures. The safest measure thus is to carefully choose immiscible metal and ceramic phases. A good example for such a combination is represented by the ceramic-metal composite of TiO_2 and Cu, the mixtures of which were found immiscible over large areas in the Gibbs' triangle of Cu-O-Ti.

The concept of ceramic-matrix nanocomposites was also applied to thin films that are solid layers of a few nm to some tens of μm thickness deposited upon an underlying substrate and that play an important role in the functionalization of technical surfaces. Gas flow sputtering by the hollow cathode technique turned out as a rather effective technique for the preparation of nanocomposite layers. The process operates as a vacuum-based deposition technique and is associated with high deposition rates up to some μm/s and the growth of nanoparticles in the gas phase. Nanocomposite layers in the ceramics range of composition were prepared from TiO_2 and Cu by the hollow cathode technique that showed a high mechanical hardness, small coefficients of friction and a high resistance to corrosion.

Metal-matrix nanocomposites

Another kind of nanocomposite is the energetic nanocomposite, generally as a hybrid sol–gel with a silica base, which, when combined with metal oxides and nano-scale aluminium powder, can form *superthermite* materials.

Polymer-matrix nanocomposites

Definitions: In the simplest case, appropriately adding nanoparticulates to a polymer matrix can enhance its performance, often in very dramatic degree, by simply capitalizing on the nature and properties of the nanoscale filler (these materials are better described by the term *nanofilled polymer composites*). This strategy is particularly effective in yielding high performance composites, when good dispersion of the filler is achieved and the properties of the nanoscale filler are substantially different or better than those of the matrix, for example, reinforcing a polymer matrix by much stiffer nanoparticles of ceramics, clays, or carbon nanotubes. Alternatively, the enhanced properties of high performance nanocomposites may be mainly due to the high aspect ratio and/or the high surface area of the fillers, since nanoparticulates have extremely high surface area to volume ratios when good dispersion is achieved.

Nanoscale dispersion of filler or controlled nanostructures in the composite can introduce new physical properties and novel behaviours that are absent in the unfilled matrices, effectively changing the nature of the original matrix (such composite materials can be better described by the term *genuine nanocomposites* or *hybrids*). Some examples of such new properties are fire resistance or flame retardant and accelerated biodegradability.

Nanoparticle Applications

Nanoparticles are particles that have one dimension that is 100 nanometers or less in size. The properties of many conventional materials change when formed from nanoparticles. This is typically because nanoparticles have a greater surface area per weight than larger particles; they are therefore more reactive to certain other molecules.

Nanoparticles are used in many fields; the list below introduces many of those uses.

- Palladium nanoparticles used in chemical vapor sensors to detect hydrogen gas.
- Quantum Dots (crystalline nanoparticles) that identify the location of cancer cells in the body.
- Iron nanoparticles used to clean up carbon tetrachloride pollution in ground water
- Silicate nanoparticles used to provide a barrier to gasses (for example oxygen), or moisture in a plastic film used for packaging. This could reduce the possibly of food spoiling or drying out.
- Zinc oxide nanoparticles dispersed in industrial coatings to protect wood, plastic and textiles from exposure to UV rays.

- Silicon dioxide crystalline nanoparticles filling gaps between carbon fibers strengthen tennis racquets.
- Silver nanoparticles in fabric that kills bacteria making clothing odor-resistant.
- Titanium oxide nanoparticles used as a photocatalyst to remove germs and other pollutants from air.
- Manganese oxide nanoparticles used as a catalyst for removal of volatile organic compounds in industrial air emissions.
- Zinc oxide nano-wires used as detection elements in sensors capable of detecting a range of chemical vapors.

❑❑❑

Chapter-3

Introduction to Semiconductors and Quantum Dots

A semiconductor is a material that has an electrical resistivity between that of a conductor and an insulator.

A semiconductor is a substance, usually a solid chemical element or compound that can conduct electricity under some conditions but not others, making it a good medium for the control of electrical current. Its conductance varies depending on the current or voltage applied to a control electrode, or on the intensity of irradiation by infrared (IR), visible light, ultraviolet (UV), or X rays.

An external electrical field changes a semiconductor's resistivity. Devices made from semiconductor materials are the foundation of modern electronics, including radio, computers, telephones, and many other devices. Semiconductor devices include the transistor, solar cells, many kinds of diodes including the light-emitting diode, the silicon controlled rectifier, and digital and analog integrated circuits. Solar photovoltaic panels are large semiconductor devices that directly convert light energy into electrical energy. In a metallic conductor, current is carried by the flow of electrons. In semiconductors, current can be carried either by the flow of electrons or by the flow of positively-charged "holes" in the electron structure of the material.

The specific properties of a semiconductor depend on the impurities, or *dopants*, added to it. An *N-type* semiconductor carries current mainly in the form of negatively-charged electrons, in a manner similar to the conduction of current in a wire. A *P-type* semiconductor carries current predominantly as electron deficiencies called holes. A hole has a positive electric charge, equal and opposite to the charge on an electron. In a semiconductor material, the flow of holes occurs in a direction opposite to the flow of electrons.

Elemental semiconductors include antimony, arsenic, boron, carbon, germanium, selenium, silicon, sulfur, and tellurium. Silicon is the best-known of these, forming the basis of most integrated circuits (ICs). Common semiconductor compounds include gallium arsenide, indium antimonide, and the oxides of most metals. Of these, gallium arsenide (GaAs) is widely used in low-noise, high-gain, and weak-signal amplifying devices.

Silicon is used to create most semiconductors commercially. Dozens of other materials are used, including germanium, gallium arsenide, and silicon carbide. A pure semiconductor is often called an "intrinsic" semiconductor. The conductivity, or ability to conduct, of semiconductor material can be drastically changed by adding other elements, called "impurities" to the melted intrinsic material and then allowing the melt to solidify into a new and different crystal. This process is called "doping".

Energy bands and electrical conduction

In solid-state physics, the electronic band structure (or simply band structure) of a solid describes ranges of energy that an electron is "forbidden" or "allowed" to have. It is due to the diffraction of the quantum mechanical electron waves in the periodic crystal lattice. The band structure of a material determines several characteristics, in particular its electronic and optical properties.

Why bands occur in materials?

The electrons of a single isolated atom occupy atomic orbitals, which form a discrete set of energy levels. If several atoms are brought together into a molecule, their atomic orbitals split, as in a coupled oscillation. This produces a number of molecular orbitals proportional to the number of atoms. When a large number of atoms (of order 10^{20} or more) are brought together to form a solid, the number of orbitals becomes exceedingly large, and the difference in energy between them becomes very small, so the levels may be considered to form continuous bands of energy rather than the discrete energy levels of the atoms in isolation. However, some intervals of energy contain no orbitals, no matter how many atoms are aggregated, forming band gaps.

Within an energy band, energy levels are so numerous as to be a near continuum. First, the separation between energy levels in a solid is comparable with the energy that electrons constantly exchange with phonons (atomic vibrations). Second, it is comparable with the energy uncertainty due to the Heisenberg uncertainty principle, for reasonably long intervals of time. As a result, the separation between energy levels is of no consequence.

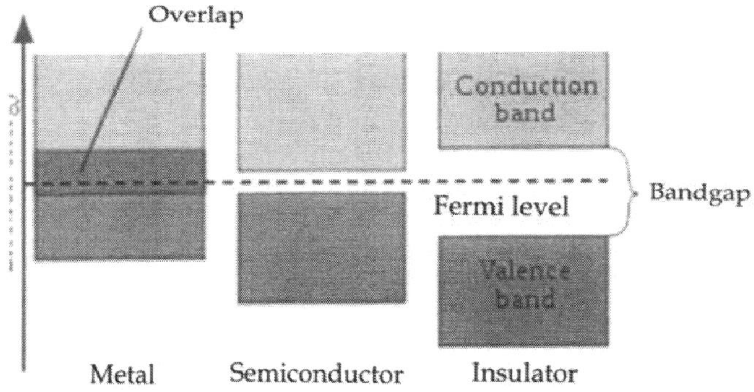

Fig. 3.1 : Simplified diagram of the electronic band structure of metals, semiconductors, and insulators.

Simplified diagram of the electronic band structure of metals, semiconductors, and insulators. Any solid has a large number of bands. In theory, it can be said to have infinitely many bands (just as an atom has infinitely many energy levels). However, all but a few lie at energies so high that any electron that reaches those energies escapes from the solid. These bands are usually disregarded.

Bands have different widths, based upon the properties of the atomic orbitals from which they arise. Also, allowed bands may overlap, producing (for practical purposes) a single large band.

Metals contain a band that is partly empty and partly filled regardless of temperature. Therefore they have very high conductivity.

The lowermost, almost fully occupied band in an insulator or semiconductor, is called the valence band by analogy with the valence electrons of individual atoms. The uppermost, almost unoccupied band is called the conduction band because only when electrons are excited to the conduction band can current flow in these materials. The difference between insulators and semiconductors is only that the forbidden band gap between the valence band and conduction band is larger in an insulator, so that fewer electrons are found there and the electrical conductivity is lower. Because one of the main mechanisms for electrons to be excited to the conduction band is due to thermal energy, the conductivity of semiconductors is strongly dependent on the temperature of the material.

This band gap is one of the most useful aspects of the band structure, as it strongly influences the electrical and optical properties of the material. Electrons can transfer from one band to the other by means of carrier generation and recombination processes. The band gap and defect states created in the band gap by doping can be used to create semiconductor devices such as solar cells, diodes, transistors, laser diodes, and others.

Like in other solids, the electrons in semiconductors can have energies only within certain bands (i.e. ranges of levels of energy) between the energy of the ground state, corresponding to electrons tightly bound to the atomic nuclei of the material, and the free electron energy, which is the energy required for an electron to escape entirely from the material. The energy bands each correspond to a large number of discrete quantum states of the electrons, and most of the states with low energy (closer to the nucleus) are full, up to a particular band called the *valence band*. Semiconductors and insulators are distinguished from metals because the valence band in the semiconductor materials is very nearly full under usual operating conditions, thus causing more electrons to be available in the "conduction band," which is the band immediately above the valence band.

The ease with which electrons in a semiconductor can be excited from the valence band to the conduction band depends on the band gap between the bands, and it is the size of this energy band gap that serves as an arbitrary dividing line between semiconductors and insulators.

Electrons excited to the conduction band also leave behind electron holes, or unoccupied states in the valence band. Both the conduction band electrons and the valence band holes contribute to electrical conductivity. The holes themselves don't actually move, but a neighboring electron can move to fill the hole, leaving a hole at the place it has just come from, and in this way the holes appear to move, and the holes behave as if they were actual positively charged particles.

Carrier generation and recombination

When ionizing radiation strikes a semiconductor, it may excite an electron out of its energy level and consequently leave a hole. This process is known as *electron–hole pair generation*. Electron-hole pairs are constantly generated from thermal energy as well, in the absence of any external energy source.

Electron-hole pairs are also apt to recombine. Conservation of energy demands that these recombination events, in which an electron loses an amount of energy larger than the band gap, be accompanied by the emission of thermal energy (in the form of phonons) or radiation (in the form of photons).

In some states, the generation and recombination of electron–hole pairs are in equipoise. The number of electron-hole pairs in the steady state at a given temperature is determined by quantum statistical mechanics. The precise quantum mechanical mechanisms of generation and recombination are governed by conservation of energy and conservation of momentum.

As the probability that electrons and holes meet together is proportional to the product of their amounts, the product is in steady state nearly constant at a given temperature, providing that there is no significant electric field (which might "flush" carriers of both types, or move them from neighbour regions containing more of them to meet together) or externally driven pair generation. The product is a function of the temperature, as the probability of getting enough thermal energy to produce a pair increases with temperature, being approximately 1×exp("E_G / kT), where k is Boltzmann's constant, T is absolute temperature and E_G is band gap.

The probability of meeting is increased by carrier traps – impurities or dislocations which can trap an electron or hole and hold it until a pair is completed. Such carrier traps are sometimes purposely added to reduce the time needed to reach the steady state.

Semi-insulators

Some materials are classified as semi-insulators. These have electrical conductivity nearer to that of electrical insulators. Semi-insulators find niche applications in micro-electronics. An example of a common semi-insulator is Gallium Arsenide.

Doping

The property of semiconductors that makes them most useful for constructing electronic devices is that their conductivity may easily be modified by introducing impurities into their crystal lattice. The process of adding controlled impurities to a semiconductor is known as *doping*. The amount of impurity, or dopant, added to an *intrinsic* (pure) semiconductor varies its level of conductivity. Doped semiconductors are often referred to as *extrinsic*. By adding impurity to pure semiconductors, the electrical conductivity may be varied not only by the number of impurity atoms but also, by the type of impurity atom and the changes may be thousand folds and million folds. For example - 1 cm^3 of a metal or semiconductor specimen has a number of atoms on the order of 10^{22}. Since every atom in metal donates at least one free electron for conduction in metal, 1 cm^3 of metal contains free electrons on the order of 10^{22}. At the temperature close to 20 °C , 1 cm^3 of pure germanium contains about 4.2×10^{22} atoms and 2.5×10^{13} free electrons and 2.5×10^{13} holes (empty spaces in crystal lattice having positive charge) The addition of 0.001% of arsenic (an impurity) donates an extra 10^{17} free electrons in the same volume and the electrical conductivity increases about 10,000 times.

Dopants

The materials chosen as suitable dopants depend on the atomic properties of both the dopant and the material to be doped. In general, dopants that produce the desired controlled changes are classified as either electron acceptors or donors. A donor atom that activates (that is, becomes incorporated into the crystal lattice) donates weakly-bound valence electrons to the material, creating excess negative charge carriers. These weakly-bound electrons can move about in the crystal lattice relatively freely and can facilitate conduction in the presence of an electric field. (The donor atoms introduce some states under, but very close to the conduction band edge. Electrons at these states can be easily excited to the conduction band, becoming free electrons, at room temperature.) Conversely, an activated acceptor produces a hole.

Semiconductors doped with donor impurities are called *n-type*, while those doped with acceptor impurities are known as *p-type*.

There are two types of impurities:

You can change the behavior of silicon and turn it into a conductor by **doping** it. In doping, you mix a small amount of an **impurity** into the silicon crystal.

N-type - In N-type doping, phosphorus or arsenic is added to the silicon in small quantities. Phosphorus and arsenic each have five outer electrons, so they're out of place when they get into the silicon lattice. The fifth electron has nothing to bond to, so it's free to move around. It takes only a very small quantity of the impurity to create enough free electrons to allow an electric current to flow through the silicon. N-type silicon is a good conductor. Electrons have a negative charge, hence the name N-type.

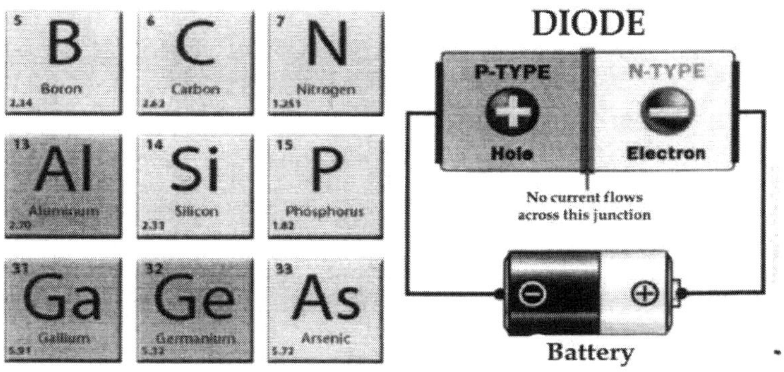

Fig. 3.2 : Semiconductor doping **Fig. 3.3** : Diode

P-type - In P-type doping, boron or gallium is the dopant. Boron and gallium each have only three outer electrons. When mixed into the silicon lattice, they form "holes" in the lattice where a silicon electron has nothing to bond to. The absence of an electron creates the effect of a positive charge, hence the name P-type. Holes can conduct current. A hole happily accepts an electron from a neighbor, moving the hole over a space. P-type silicon is a good conductor.

A minute amount of either N-type or P-type doping turns a silicon crystal from a good insulator into a viable (but not great) conductor — hence the name "semiconductor."

N-type and P-type silicon are not that amazing by themselves; but when you put them together, you get some very interesting behavior at the junction. That's what happens in a diode.

A *diode* is the simplest possible semiconductor device. A diode allows current to flow in one direction but not the other. You may have seen turnstiles at a stadium or a subway station that let people go through in only one direction. A diode is a one-way turnstile for electrons.

Even though N-type silicon by itself is a conductor, and P-type silicon by itself is also a conductor, the combination does not conduct any electricity. The negative electrons in the N-type silicon get attracted to the positive terminal of the battery. The positive holes in the P-type silicon get attracted to the negative terminal of the battery. No current flows across the junction because the holes and the electrons are each moving in the wrong direction.

If you *flip the battery around*, the diode conducts electricity just fine. The free electrons in the N-type silicon are repelled by the negative terminal of the battery. The holes in the P-type silicon are repelled by the positive terminal. At the **junction** between the N-type and P-type silicon, holes and free electrons meet. The electrons fill the holes. Those holes and free electrons cease to exist, and new holes and electrons spring up to take their place. The effect is that **current flows** through the junction.

The n and p type designations indicate which charge carrier acts as the material's majority carrier. The opposite carrier is called the minority carrier, which exists due to thermal excitation at a much lower concentration compared to the majority carrier.

For example, the pure semiconductor silicon has four valence electrons. In silicon, the most common dopants are IUPAC group 13 (commonly known as *group III*) and group 15 (commonly known as *group V*) elements. Group 13 elements all contain three valence electrons, causing them to function as acceptors when used to dope silicon. Group 15 elements have five valence electrons, which allows them to act as a donor. Therefore, a silicon crystal doped with boron creates a p-type semiconductor whereas one doped with phosphorus results in an n-type material.

How Semiconductors Work

Semiconductors have had a monumental impact on our society. You find semiconductors at the heart of microprocessor chips as well as transistors. Anything that's computerized or uses radio waves depends on semiconductors. Today, most semiconductor chips and transistors are created with *silicon.* You may have heard expressions like "Silicon Valley" and the "silicon economy," and that's why — silicon is the heart of any electronic device.

A *diode* is the simplest possible semiconductor device, and is therefore an excellent beginning point if you want to understand how semiconductors work.

Silicon is a very common element — for example, it is the main element in sand and quartz. If you look "silicon" up in the periodic table, you will find that it sits next to aluminum, below carbon and above germanium.

Carbon, silicon and germanium (germanium, like silicon, is also a semiconductor) have a unique property in their electron structure — each has **four electrons in its outer orbital**. This allows them to form nice crystals. The four electrons form perfect covalent bonds with four neighboring atoms, creating a **lattice.** In carbon, we know the crystalline form as diamond. In silicon, the crystalline form is a silvery, metallic-looking substance.

- Metals tend to be good conductors of electricity because they usually have "free electrons" that can move easily between atoms, and electricity involves the flow of electrons. While silicon crystals look metallic, they are not, in fact, metals. All of the outer electrons in a silicon crystal are involved in **perfect covalent bonds**, so they can't move around. A pure silicon crystal is nearly an **insulator** — very little electricity will flow through it. But you can change all this through a process called doping.

Quantum Dots

A quantum dot is a particle of matter so small that the addition or removal of an electron changes its properties in some useful way. All atoms are, of course, quantum dots, but multi-molecular combinations can have this characteristic. In biochemistry, quantum dots are called redox groups. In nanotechnology, they are called quantum bits or qubits. Quantum dots typically have dimensions measured in nanometers.

The use of semiconductors has greatly increased in the last century. Traditional semi-conductor devices have been found to be too big and too slow. As engineers search for a faster and more adaptable alternative to conventional semiconductors they have discovered quantum dots, a new form of semiconductors that model atoms. Being only nanometers in size, these pseudo-atoms take semi-conductors to a whole new level and can allow devices to work almost at the speed of light. Furthermore, quantum dots have numerous applications in optical technologies, mediums, and industries. This section seeks to introduce the principle of quantum dots, their creation methods, and their applications.

Quantum dots (QDs) are particles that hold a droplet of free electrons which simulates "the ultimate miniaturized semiconductor." Any material that can conduct electricity better than an insulator but not as well as a conductor is considered a semi-conductor. What makes semi-conductors so important is that their unique structure allows different semi-conductors to carry current under different circumstances. This gives the user more control over the flow of current. Most semi-conductors are crystalline substances such as germanium and silicon.

Conventional semi-conductors are used often in electrical circuits. However, they have limited ranges of tolerance for the frequency of the current they carry. The low tolerance of traditional semi-conductors often poses a problem to circuits, and many of its other applications. This is what makes the use of quantum dots so important. As they are fabricated artificially, different quantum dots can be made to tolerate different current frequencies through a much larger range than conventional ones (Figure 3.4). The use of quantum dots as semi-

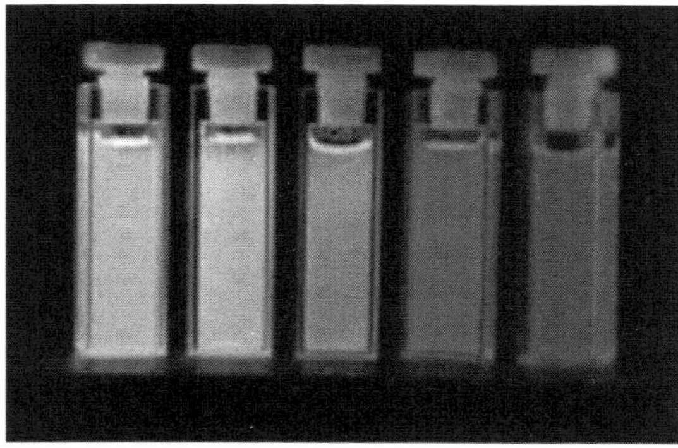

Fig. 3.4 : White light is shined on vials containing a solution that holds quantum dots engineered for different frequencies.

conductors offers more freedom to just about everything involving the use of semi-conductors (Quantum Dots Explained, 2005).

The alteration of the band-gap makes each vial absorb and re-emit a different wavelength of light or in other words each vial of quantum is engineered to show a different color of light.

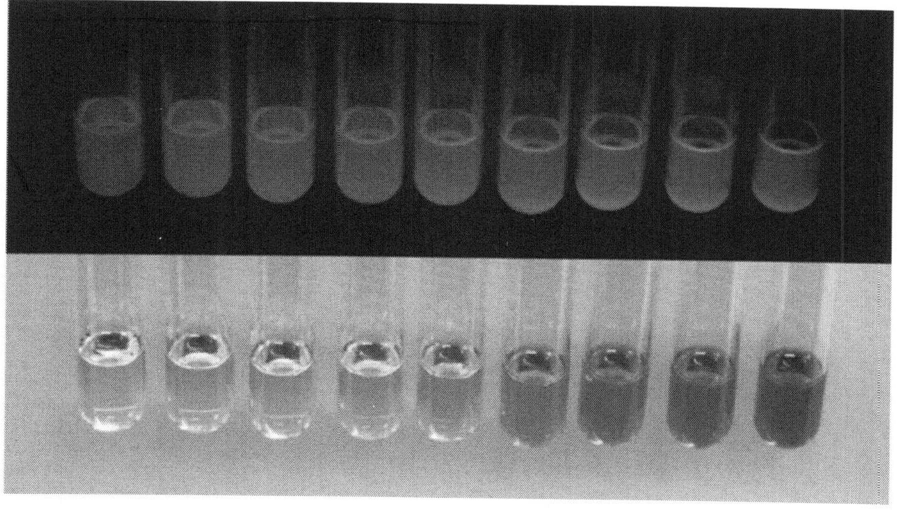

Fig. 3.5 : Preparation of Cd Se Quantum Dots

Quantum dots can best be described as false atoms. The primary material that a quantum dot is made out of is called a "hole", or a substance that is missing an electron from its valence band giving it a positive charge. The primary material is extremely small, which is why it is called a dot, and at that size, electrons start to orbit it. Since quantum dots do not have protons or neutrons in the center, their mass is much smaller. Since the mass at the center is smaller than that of an atom, quantum dots exert a smaller force on the orbiting electrons causing an orbit larger than that of a regular atom (Figure 3.5). With a mass that small, scientists are able to precisely calculate and change the size of the band-gap of the quantum dot by adding or taking electrons. The band-gap of a quantum dot is what determines which frequencies it will respond to, so being able to change the band-gap is what gives scientists more control and more flexibility when dealing with its applications (Quantum Dots Explained, 2005).

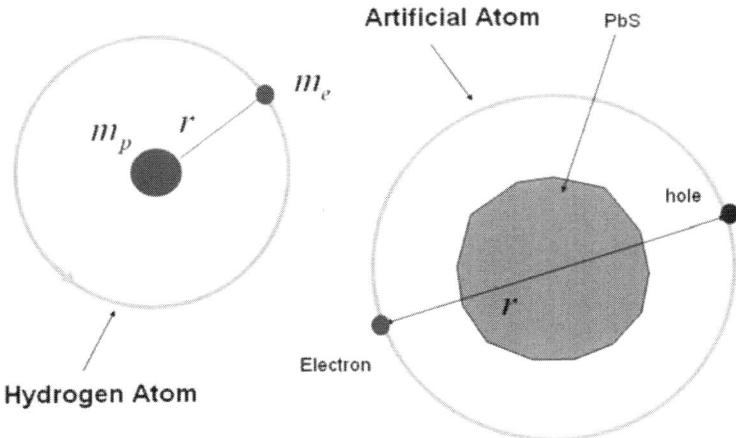

Fig. 3.5 (A) : Comparison of a hydrogen atom to a quantum dot.

The above image compares the orbit of a hydrogen atom to that of a quantum dot. Since the artificial atom has almost no mass compared to the hydrogen atom, the orbit is much larger.

One way to synthesize quantum dots is through molecular beam epitaxy. In this process, certain chemicals are evaporated and then sprayed to condense into small objects on a substrate (Molecular Beam Epitaxy, 2005). The condensation of the chemical on the substrate is similar to water on glass. If someone drops water on glass the water condenses into many balls (Figure 3.6). As more layers are sprayed onto the substrate the size of the balls starts to build up into pyramid-shaped

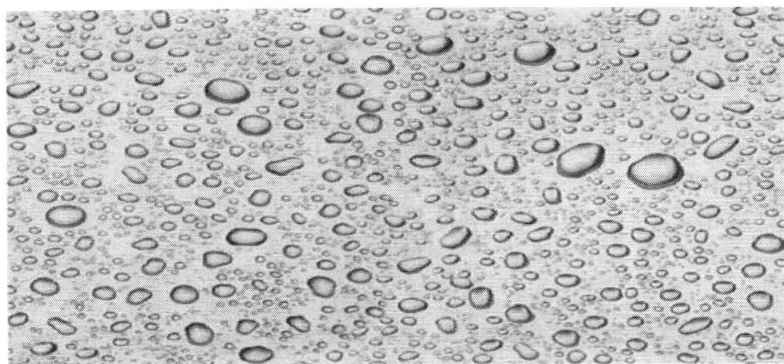

Fig. 3.6 : The different chemicals, once sprayed onto the substrate, acts almost like water and form together in "balls", or quantum dots.

objects. Eventually, the balls build up to a specific size and they're quantum dots. This process has some downsides though it is much harder to use quantum dots while they are still attached to the substrate. While they are all attached together on the substrate they act as one solid which almost defeats the purpose of creating the quantum dots.

Another way to form quantum dots is through electron beam lithography. This process is a little like etching a chip. A mask is created with an electron beam that has many tiny holes in it. Then evaporated chemicals, similar to the ones used in epitaxy, are sprayed through the mask onto a substrate, creating many little balls (Figure 3.7). This process has some of the same shortcomings as epitaxy, mainly that the quantum dots are still connected to the substrate after synthesis. Additionally, scientists have found it difficult to create such small masks that need to have holes just nanometers in diameter. Lithography was originally a very popular process for creating quantum dots; however, this process creates many defects and is slow compared to the other processes (Electron Beam Lithography, 2005).

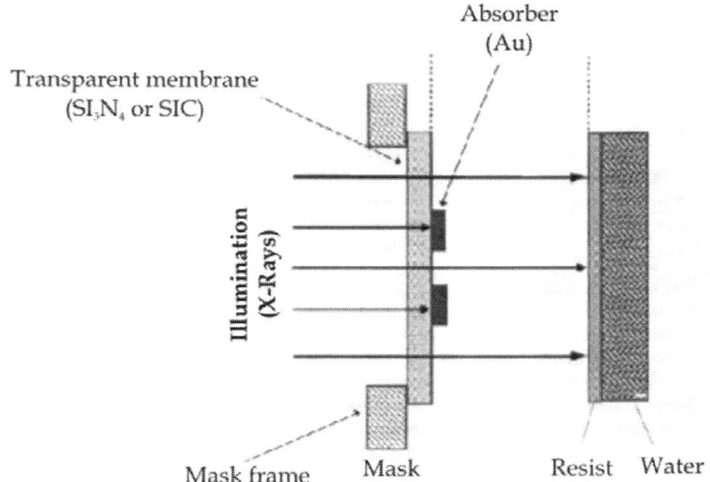

Fig. 3.7 : The illustration shows x-rays being shined through a mask. The x-ray light reacts with the photo-resist on the wafer to create quantum dots.

Colloidal synthesis is a process that involves creating quantum dots in a liquid. This is by far the best technique for the formation of quantum dots because the process can occur under "benchtop conditions," or in a normal laboratory setting. When certain materials, primarily those from periodic groups two through four, are dissolved in a certain type of polymer solute the solution can enter into a phase where particles can come together to form quantum dots. Since the size is dependent on time, the longer the dots are left in the solution the bigger they get. This is in part what makes colloidal synthesis the most popular method. Scientists can use time to change the properties of the quantum dot, engineering it for certain light frequencies. This process, unlike lithography and epitaxy, synthesizes the quantum dots in such a way that they are suspended individually, making it easier for use in applications (Colloidal Particles, 2005)

As mentioned, QDs have many interesting characteristics that not only contribute to numerous applications but also display phenomena that are consequential to the fundamental study of Physics. Typical dimensions of QDs are between nanometers to a few microns (Quantum dot, 2004). Due to their extremely small size, QDs can be controlled by a few electrons and this provides many advantages that can optimize devices. In addition to the nanoscale size that they exhibit, QDs enable superior transport and optical properties that has beneficial applications in the fields including immuno cytochemistry and the study of lasers.

In the biological field of science, QDs have become known to be very useful. Recent studies of QDs have resulted in developing new fluorescence immunocytochemical probes. A probe is a substance that is radioactively labeled or otherwise marked and used to detect or identify another substance in a sample. A fluorescence immunocytochemical probe is usually used to detect antigens in tissues (Figure 3.8). In contrast to organic fluorophores, which are not

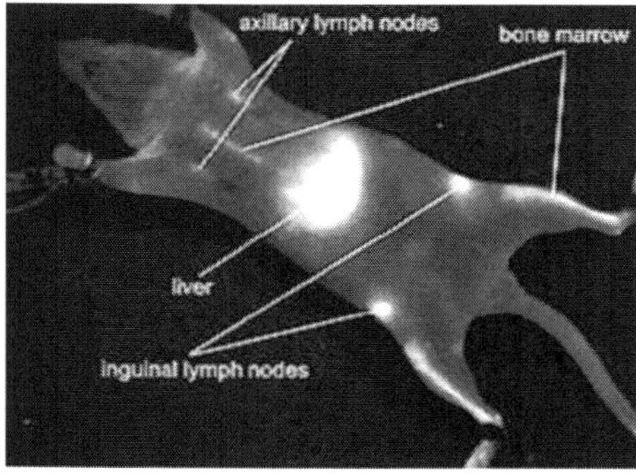

Fig. 3.8 : QD as fluoresceace immunocy to chemical probe

photostable, QDs have properties of high brightness, photostability, narrow emission spectra and an apparent large Stokes' shift, thus they can replace the usage of organic fluorophores. The current mode of detecting the antigens which takes from two to six days can speed up to a matter of hours using quantum dots.

Immunocytochemical probes are used in the dead rodent. The probes in the body have circulated and now show up under florescent light, creating a much safer alternative to the x-ray.

Prior to the introduction of the QD, microelectronic technology has focused on reducing the size of transistors to produce increasingly smaller, faster and more efficient computers. However, this method is reaching its physical limit due to the restrictions placed by the laws of physics that do not allow these devices to operate below a certain size. With this advantageous feature of QDs, information storage can be brought down to the molecular level. Since no flow of electrons to transmit a signal is needed, electric current does not need to be produced and heat problems are avoided. Also, the quantum dot devices are sensitive enough to and can make a usage of the charges of single electrons. With improvements in quantum-dot ordering and positioning, it is possible for us to hope in the near future to address and store information optically in a single quantum dot, thus opening the possibility of ultrahigh-density memory devices.

QDs also have other applications like quantum dot lasers which promises far more great advantages than quantum well lasers. Because QD lasers are less temperature-dependent and less likely to degrade under elevated temperature, it allows more flexibility for lasers to operate more efficiently. Other beneficial features of QD lasers include low threshold currents, higher power, and great stability compared to the restrained performance of the conventional lasers. Respectively, the QD laser will play a significant role in optical data communications and optical networks.

Optical switches have been a major research objective in the scientific community. The use of optical switches would increase the rate at which data can be transferred. With regular switches, data can only travel as fast as the electrical current can, but optical switches can travel almost as fast as the speed of light. The principle of optical switches is that semi-conductors will only allow certain levels of energy to pass through it. So if we place a quantum dot semi-conductor in a circuit, but supply a voltage below the acceptable range, current will not flow. However, if we shine a light on the quantum dot semi-conductor, it would put enough energy into the semi-conductor that it will allow current to flow (Figure 3.9). This idea is mainly for powering electronic devices, but using quantum-dots as receivers for electrical data is just a step up.

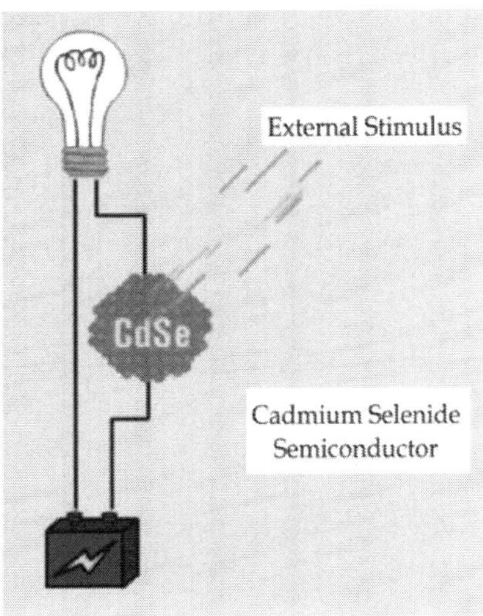

Fig. 3.9 : Optical Switch

The added stimulant to the circuit puts enough energy through the quantum dot to allow the flow of current.

Quantum dots have applications outside of biology and engineering. An idea that may be instituted in the future would be against counterfeiting money. The treasury could engineer quantum dots to be responsive to a specific frequency of light and suspend them in ink that they would print onto money. Shining light with the same frequency that the ink solution has been engineered for would reveal whether or not the money is real or counterfeit. This idea can be used for just about any substance that could be illegally duplicated.

Another security application involves attaching quantum dots to dust. QDs can be engineered so that they have the same properties as dust and give off infrared radiation. In hostile areas, this "quantum dust" can be used to track wanted criminals or the movement of hostile activity. In urban areas, "quantum dust" can be used as a security device to set off alarms if the infrared radiation is detected.

Though QDs are still under research for other possible applications and need more technological advancement in order to be put into use, the features introduced will grant far better optical communication, significant change in electronic devices, and even detection of antigens in the body tissues.

How Quantum Dots Work

A Special Class of Semiconductors

Quantum dots, also known as nanocrystals, are a special class of materials known as semiconductors, which are crystals composed of periodic groups of II-VI, III-V, or IV-VI materials. Semiconductors are a cornerstone of the modern electronics industry and make possible applications such as the Light Emitting Diode and personal computer. Semiconductors derive their great importance from the fact that their electrical conductivity can be greatly altered via an external stimulus (voltage, photon flux, etc), making semiconductors critical parts of many different kinds of electrical circuits and optical applications. Quantum dots are unique class of semiconductor because they are so small, ranging from 2-10 nanometers (10-50 atoms) in diameter. At these small sizes materials behave differently, giving quantum dots unprecedented tenability and enabling never before seen applications to science and technology.

The usefulness of quantum dots comes from their peak emission frequency's extreme sensitivity to both the dot's size and composition, which can be controlled using Evident Technologies' proprietary engineering techniques. This remarkable sensitivity is quantum mechanical in nature, and is explained as follows.

Bands and Bandgaps

The electrons in bulk (much bigger than 10 nm) semiconductor material have a range of energies. One electron with a different energy than a second electron is described as being in a different energy level, and it is established that only two electrons can fit in any given level. In bulk, energy levels are very close together, so close that they are described as continuous, meaning there is almost no energy difference between them. It is also established that some energy levels are simply off limits to electrons; this region of forbidden electron energies is called the bandgap, and it is different for each bulk material. Electrons occupying energy levels below the bandgap are described as being in the valence band. Electrons occupying energy levels above the bandgap are described as being in the conduction band.

Electrons and Holes

In natural bulk semiconductor material, an extremely small percentage of electrons occupy the conduction band the overwhelming majority of electrons occupy the valence band, filling it almost completely. The only way for an electron in the valence band to jump to the conduction band is to acquire enough energy to cross the bandgap, and most electrons in bulk simply do not have enough energy to do so. Applying a stimulus such as heat, voltage, or photon flux can induce some electrons to jump the forbidden gap to the conduction band. The valence location they vacate is referred to as a hole since it leaves a temporary "hole" in the valence band electron structure.

Bulk Semiconductors - A Fixed Range of Energies

A sufficiently strong stimulus will cause a valence band electron to take residence in the conduction band, causing the creation of a positively charged hole in the valence band. The raised electron and the hole taken as a pair are called an exciton. There is a minimum energy of radiation that the semiconductor bulk can absorb towards raising electrons into the conduction band, corresponding to the energy of the bandgap. It is established that because of the continuous electron energy levels as well as the number of atoms in the bulk, the bandgap energy of bulk semiconductor material of a given composition is fixed.

It is also established that electrons in natural semiconductor bulk that have been raised into the conduction band will stay there only momentarily before falling back across the bandgap to their natural, valence energy levels. As the electron falls back down across the bandgap, electromagnetic radiation with a wavelength corresponding to the energy it loses in the transition is emitted. It is established that the great majority of electrons, when falling from the conduction band back to the valence band, tend to jump from near the bottom of the conduction band to the top of the valence band- in other words, they travel from one edge of the bandgap to the other. Because the bandgap of the bulk is fixed, this transition results in fixed emission frequencies. Quantum dots offer the unnatural ability to tune the bandgap and hence the emission wavelength.

Quantum Dots - Quantum Confinement

Quantum dots are also made out of semiconductor material. The electrons in quantum dots have a range of energies. The concepts of energy levels, bandgap, conduction band and valence band still apply. However, there is a major difference. Excitons have an average physical separation between the electron and hole, referred to as the Exciton Bohr Radius. This physical distance is different for each material. In bulk, the dimensions of the semiconductor crystal are much larger than the Exciton Bohr Radius, allowing the exciton to extend to its natural limit. However, if the size of a semiconductor crystal becomes small enough that it approaches the size of the material's Exciton Bohr Radius, then the electron energy levels can no longer be treated as continuous - they must be treated as discrete, meaning that there is a small and finite separation between energy levels. This situation of discrete energy levels is called quantum confinement, and under these conditions, the semiconductor material ceases to resemble bulk, and instead can be called a quantum dot. This has large repercussions on the absorptive and emissive behavior of the semiconductor material.

Quantum Dots - A tunable range of energies

Because quantum dots' electron energy levels are discrete rather than continuous, the addition or subtraction of just a few atoms to the quantum dot has the effect of altering the boundaries of the bandgap. Changing the geometry of the surface of the quantum dot also changes the bandgap energy, owing again to the small size of the dot, and the effects of quantum confinement. The bandgap in a quantum dot will always be energetically larger; therefore, we refer to the radiation from quantum dots to be "blue shifted" reflecting the fact that electrons must fall a greater distance in terms of energy and thus produce radiation of a shorter and therefore "bluer" wavelength.

Size Dependent Control of Bandgap in Quantum Dots

As with bulk semiconductor material, electrons tend to make transitions near the edges of the bandgap. However, with quantum dots, the size of the bandgap is controlled simply by adjusting the size of the dot. Because the emission frequency of a dot is dependent on the bandgap, it is therefore possible to control the output wavelength of a dot with extreme precision. In effect, it is possible to tune the bandgap of a dot, and therefore specify its "color" output depending on the needs of the customer.

❑❑❑

Chapter-4

Polymeric Nanoparticles

Polymeric nanoparticles are nanoparticles which are prepared from polymers. The drug is dissolved, entrapped, encapsulated or attached to nanoparticles and depending upon the method of preparation, nanoparticles, nanospheres or nanocapsules can be obtained. Nanocapsules are vesicular systems in which the drug is confined to a cavity surrounded by a polymer membrane, while nanospheres are matrix systems in which the drug is physically and uniformly dispersed.

In recent years, biodegradable polymeric nanoparticles have attracted considerable attention as potential drug delivery devices in view of their applications in drug targeting to particular organs/tissues, as carriers of DNA in gene therapy, and in their ability to deliver proteins, peptides and genes through oral route of administration.

In spite of development of various synthetic and semi synthetic polymers, natural polymers still enjoy their popularity in drug delivery, some of them are: gums (Eg. Acacia, Guar, etc.,) chitosan, gelatin, sodium alginate, and albumin. A range of materials have been employed to delivery of bioactive agents. A polymer used in controlled drug delivery formulation, must be chemically inert, non-toxic and free of leachable impurities. It must also have an appropriate physical structure, with minimal undesired aging, and be readily processable. Some of the polymeric materials are: cellulosics., poly(2-hydroxy ethyl methacrylate), poly(N-vinyl pyrrolidone), poly(methyl methacrylate), poly(vinyl alcohol), poly(acrylic acid), polyacrylamide, poly(ethylene-co-vinyl acetate), poly(ethylene glycol), poly(methacrylic acid).

However, in recent years additional polymers are designed primarily for medical applications and have entered the arena of controlled release of bioactive agents. Many of these materials are designed to degrade within the body, most popular ones are: polylactides (PLA), polyglycolides (PGA), poly(lactide-co-glycolides) (PLGA), polyanhydrides, polyorthoesters, polycyanoacrylates, polycaprolactone. Originally, polylactides and polyglycolides were used as absorbable suture material. The main advantage of these degradable polymers is that they are broken down into biologically acceptable molecules that are metabolized and removed from the body via normal metabolic pathways. However, biodegradable materials do produce degradation by-products that must be tolerated with little or no adverse reactions within the biological environment.

Abraxane is the first polymeric nanoparticle based product from American Pharmaceutical Partners Inc., and American Bioscience, Inc. (ABI). It was approved in year 2005 and consists of albumin-bound paclitaxel nanoparticles. This product is free of toxic solvents like cremophor-EL, which is used until now to solubulize paclitaxel in order to administer it intravenously to the patient. Cremophor-EL is known to cause life-threatening allergic reactions. Success of Abraxane shows that nanotechnology can bring many exciting products which can overcome many hurdle of formulation scientist

Engineering of polymeric nanoparticles

Polymeric nanoparticles are engineered by various methods and some of them are listed below;

- Emulsification followed by solvent evaporation/cross-linking
- Emulsion polymerization
- Spray drying
- Supercritical fluid technology
- Electrospray method

Organic Materials and Polymer Synthesis

Synthesis of electro-optically active materials: polymers, oligomers, amphiphiles, polyelectrolytes and dendrimers. This includes the use of the precursor polymer method. Application of electrochemistry and electropolymerization methods to synthesize ultrathin films with conjugated polymers and oligomers.

Synthesis and photo-induced alignment of dyes in ultra thin films using photoisomerizable moieties: understanding the alignment process and application for display devices. Fabrication of materials and thin films specifically for Organic Light Emitting Diodes (OLED) and Field Effect Transistors (FET) devices using ultra thin film techniques.

Development of patterning and non-conventional deposition techniques using reactive cross-linking processes.

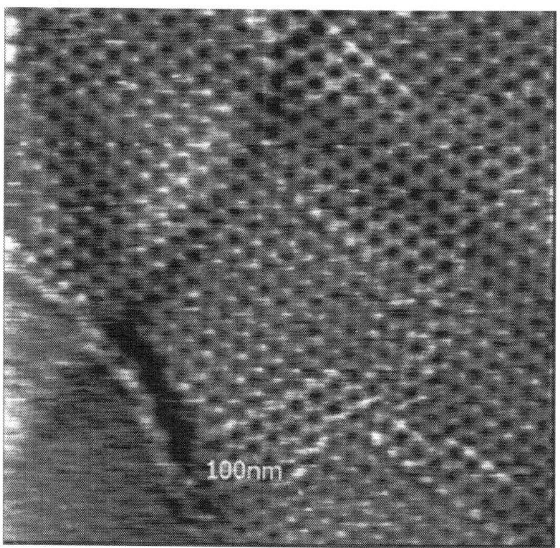

Fig. 4.1 : 3-D Crystallization of Thiophene Dendrimers

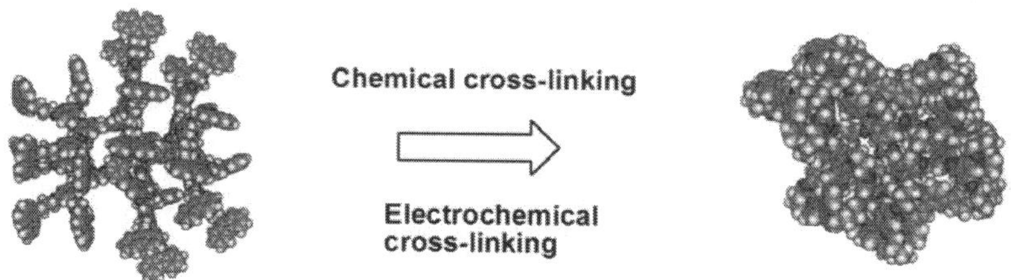

Fig. 4.2 : Conjugated Polymer Nanoparticles

Ultra thin Films and Surface Analysis

Investigation of polymer and small molecule adsorption on surfaces using surface sensitive techniques: QCM, SPS, AFM, ellipsometry, waveguides. Surface Initiated Polymerization (SIP) methods for modifying surfaces, coating of nanoparticles (core shells) and formation of nanocomposites. Application of the Electrostatic Layer-By-Layer (ELBL) methods to form functional ultrathin films. Application of electropolymerization techniques using the precursor polymers and designed oligomers.

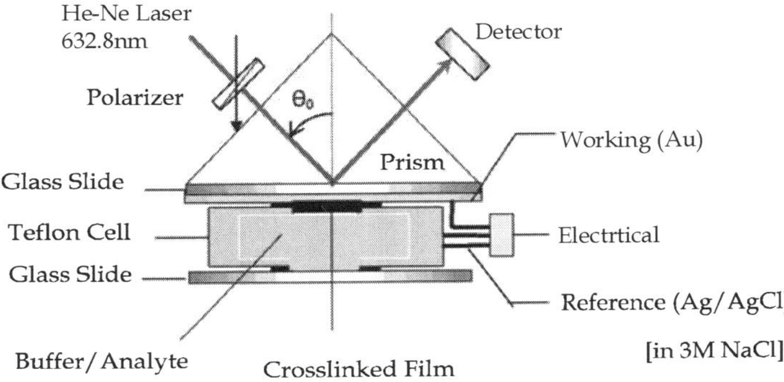

Fig. 4.3 : Electrochemical Surface Plasmon Resonance Spectroscopy Set-Up

Nanotechnology Related Projects

Synthesis of Nanoparticles using polyelectrolyte complexes and block copolymers.Nanostructured Ultrathin Films Using Layer-by-layer Assemblies to incorporate nanobuilding blocks Dendrimers as Organic Macromolecule with applications in energy transfer, electron transfer, guest-host (molecular imprinting), and core-shell systems.

Biotechnology Related Projects

Development of biofunctional surfaces and ultrathin film strategies as applied to biomedical and bioengineering: applications to drug delivery, implants, and biosensors. Supramolecular assembly and layer-by-layer techniques in ultrathin films as applied to biotechnology.

Branched Polymer Nanoparticles

The production of materials with control over structure on the nanometer scale is of fundamental importance in science and technology.

There are relatively few nonbiological examples of the direct covalent chemical synthesis of complex, non-spherical organic nanostructures with length scales beyond 10's of nanometres. We have recently demonstrated the direct, targeted, one-pot synthesis of spherical and anisotropic organic nanostructures on a multigram scale from simple vinyl monomers. Polymer nanoparticles with targeted "dumbbell" shapes were prepared without requiring any programmed self assembly steps .

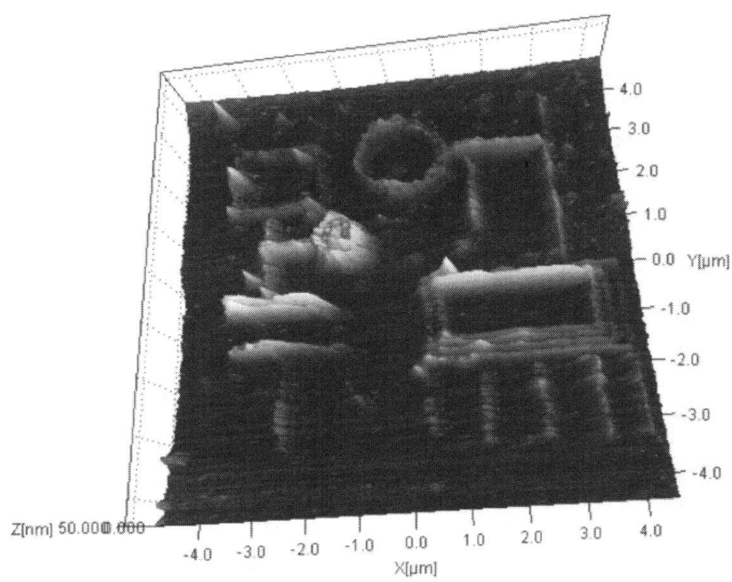

Fig. 4.4 : Electrochemical Nanopatterning of Precursor Polymers

Biopolymers

Biopolymers are a class of polymers produced by living organisms. Cellulose and starch, proteins and peptides, and DNA and RNA are all examples of biopolymers, in which the monomeric units, respectively, are sugars, amino acids, and nucleotides.

Cellulose is the most common biopolymer and the common organic compound on Earth, and about 33 percent of all plant matter is cellulose (the cellulose content of cotton is 90 percent and that of wood is 50 percent).

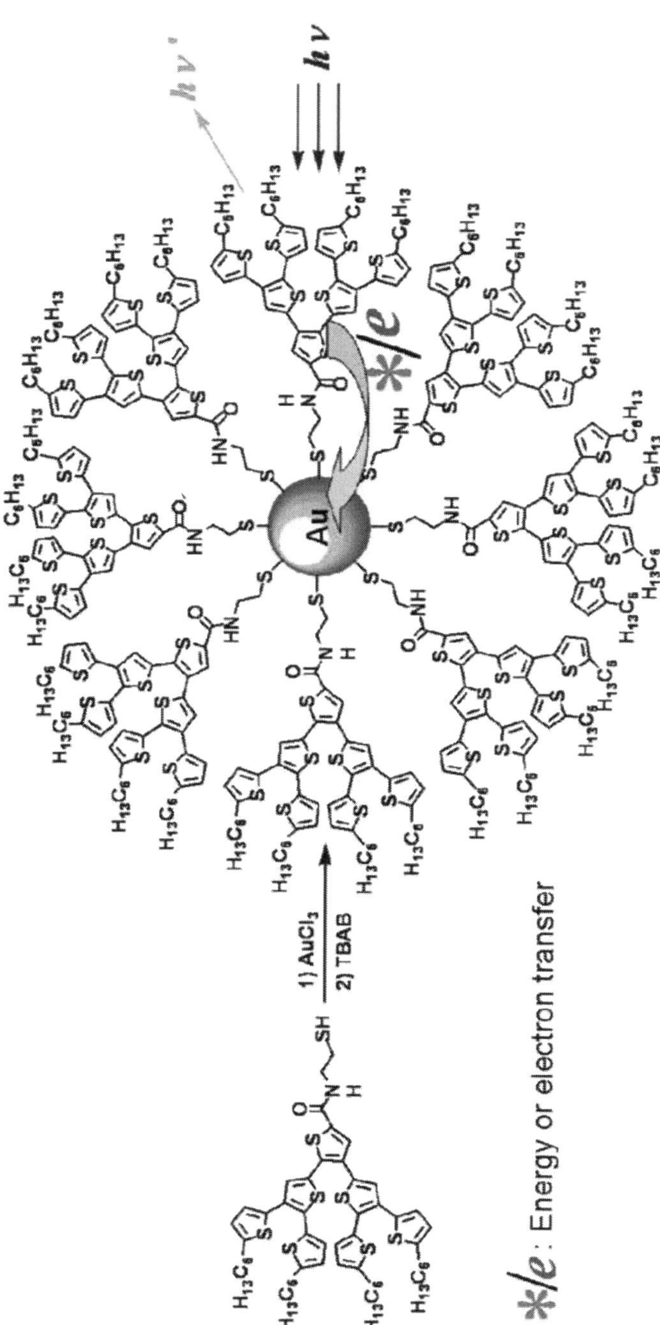

Fig. 4.5 : Hybrid-Au-Dendron Nanoparticles

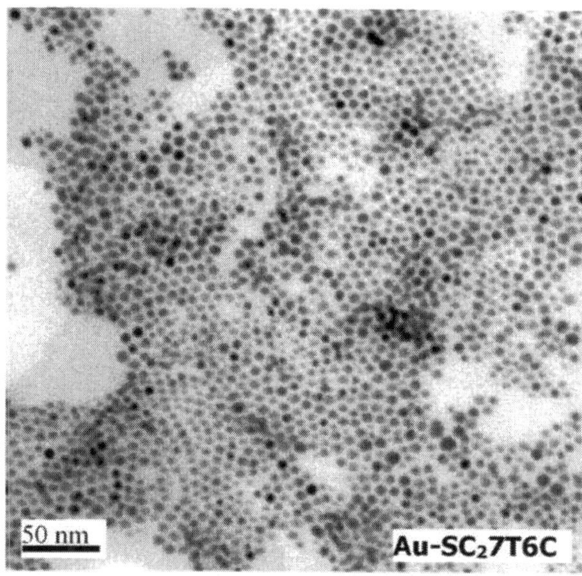

Fig. 4.6 : TEM of Hybrid Au-Thiophene Dendron Nanoparticles

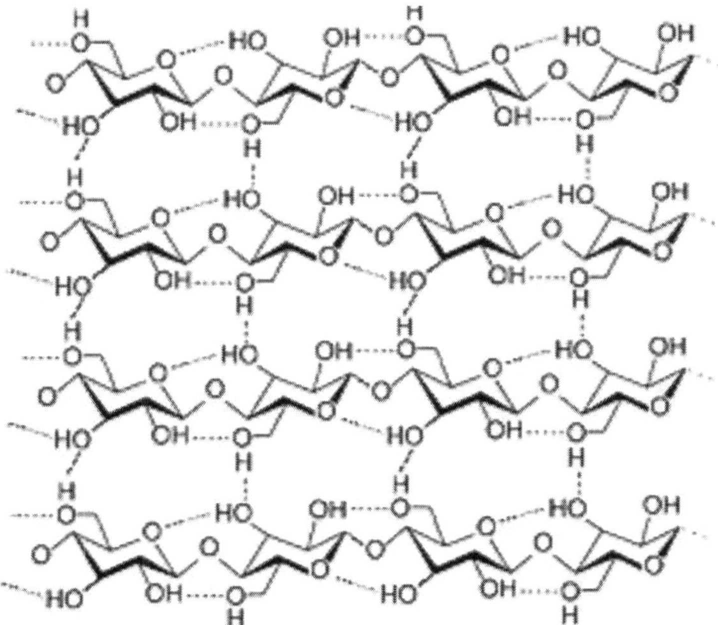

Fig. 4.7 : Cellulose is a biopolymer

Biopolymers versus polymers

A major but defining difference between polymers and biopolymers can be found in their structures. Polymers, including biopolymers, are made of repetitive units called monomers. Biopolymers often have a well defined structure, though this is not a defining characteristic (example:ligno-cellulose): The exact chemical composition and the sequence in which these units are arranged is called the primary structure, in the case of proteins. Many biopolymers spontaneously fold into characteristic compact shapes, which determine their biological functions and depend in a complicated way on their primary structures. Structural biology is the study of the structural properties of the biopolymers. In contrast most synthetic polymers have much simpler and more random (or stochastic) structures. This fact leads to a molecular mass distribution that is missing in biopolymers. In fact, as their synthesis is controlled by a template directed process in most in vivo systems all biopolymers of a type (say one specific protein) are all alike: they all contain the similar sequences and numbers of monomers and thus all have the same mass. This phenomenon is called monodispersity in contrast to the polydispersity encountered in synthetic polymers. As a result biopolymers have a polydispersity index of 1.

Structural characterization

There are a number of biophysical techniques for determining sequence information. Protein sequence can be determined by Edman degradation, in which the N-terminal residues are hydrolyzed from the chain one at a time, derivatized, and then identified. Mass spectrometer techniques can also be used. Nucleic acid sequence can be determined using gel electrophoresis and capillary electrophoresis. Lastly, mechanical properties of these biopolymers can often be measured using optical tweezers or atomic force microscopy.

Biopolymers as materials

Some biopolymers- such as polylactic acid (PLA), naturally occurring zein, and poly-3-hydroxybutyrate can be used as plastics, replacing the need for polystyrene or polyethylene based plastics.

Some plastics are now referred to as being 'degradable', 'oxy-degradable' or 'UV-degradable'. This means that they break down when exposed to light or air, but these plastics are still primarily (as much as

98 per cent) oil-based and are not currently certified as 'biodegradable' under the European Union directive on Packaging and Packaging Waste. Biopolymers, however, will break down and some are suitable for domestic composting.

Biopolymers as packaging

Biopolymers (also called renewable polymers) are produced from biomass for use in the packaging industry. Biomass comes from crops such as sugar beet, potatoes or wheat: when used to produce biopolymers, these are classified as non food crops. These can be converted in the following pathways:

Sugar beet > Glyconic acid > Polyglonic acid

Starch > (fermentation) > Lactic acid > Polylactic acid (PLA)

Biomass > (fermentation) > Bioethanol > Ethene > Polyethylene

Many types of packaging can be made from biopolymers: food trays, blown starch pellets for shipping fragile goods, thin films for wrapping.

Biopolymers are renewable, sustainable, and can be carbon neutral

Biopolymers are renewable, because they are made from plant materials which can be grown year on year indefinitely. These plant materials come from agricultural non food crops. Therefore, the use of biopolymers would create a sustainable industry. In contrast, the feedstocks for polymers derived from petrochemicals will eventually run out. In addition, biopolymers have the potential to cut carbon emissions and reduce CO_2 quantities in the atmosphere: this is because the CO_2 released when they degrade can be reabsorbed by crops grown to replace them: this makes them close to carbon neutral.

Biopolymers are biodegradable, and some are also compostable

Some biopolymers are biodegradable: they are broken down into CO_2 and water by microorganisms. In addition, some of these biodegradable biopolymers are compostable: they can be put into an industrial composting process and will break down by 90% within 6 months. Biopolymers that do this can be marked with a 'compostable' symbol, under European Standard EN 13432 (2000). Packaging marked with this symbol can be put into industrial composting processes and will break down within 6 months (or less). An example of a compostable polymer is PLA film under 20ìm thick: films which are thicker than that do not qualify as compostable, even though they are biodegradable.

❑❑❑

Chapter-5

Core / Shell Nanoparticles

"Small is beautiful", today we know that small is not only beautiful but also powerful. With the range of applications that nanoparticles find in varied fields of engineering and science, nanoparticles seem a promising option when compared to the conventional materials used. Nanoparticles are particles that have at least one dimension in less than 100nm range. They have a high surface to volume ratio and thus mass transfer and heat transfer properties are better than bulk materials. Recently; core/shell nanoparticles are finding widespread application. There is a class of core/shell nanoparticle that has its entire constituent in the nanometer range. Core/shell nanoparticles are nanostructures that have core made of a material coated with another material. They are in the size range of 20nm-200nm. Also, composite structures with these core/shell particles embedded in a matrix material are in use. The necessity to shift to core/shell nanoparticles is the improvement in the properties. Taking into consideration the size of the nanoparticles, the shell material can be chosen such that the agglomeration of particle can be prevented. This implies that the monodispersity of the particles can be improved. The core/shell structure enhances the thermal and chemical stability of the nanoparticles, improves solubility, makes them less cytotoxic and allows conjugation of other molecules to these particles. The shell can also prevent the oxidation of the core material. "When a core nanoparticle is coated with a polymeric layer or an inorganic layer like silica because the polymeric or inorganic layer would endow the hybrid structure with an additional function/property on top of the function/property of the core hence synergistically emerged functions can be envisioned".

Classification of Core/Shell Nanoparticles

The properties of the nanomaterials are diverse and cannot be generalized even though the particles under comparison might be made of similar material and com-position. This rule is applicable to core/shell nanoparticles and hence, classified them broadly based on the material with which the core and shell of the nanocomposite are made. On these lines we can group the nanoparticles as follows

1. Inorganic Core/Shell Nanoparticles

The core or the shell or both are made of inorganic materials. The inorganic materials in most commercially used and widely synthesized nanoparticles are metals, semiconductors or lanthanides.

1.1. Metallic Nanoparticles

The core or the shell should contain metallic component. The core of such structures can be a metal or metal oxide or inorganic like silica while the shell can also be any inorganic material like silica or can be a metal or metal oxide. The most widely used core/shell nancomposites are gold or silver core with silica shell. The gold/silica nanoparticles find application in optical sensing and the thickness of the silica coat alters the optical property of gold nanoparticles. The silica coat makes the gold nanoparticles biocompatible. The silver/silica particles can be used in fluorescence imaging and again the region of emission is dependent on the silica coating thickness. The structure can be inversed and silica nanoparticles can form the core and the gold can form the shell. The thickness of the gold coat can be varied and a range of colors can be obtained. These particles find application in fluorescent bioimaging. Metal interaction with the fluorophores increases photo stability, enhances fluorescence, reduces lifetime. Also fluorescent core/shell nanoparticles enhance single particle imaging. Other core/shell nanoparticles that find application in optical bioimaging are copper/copper oxide and these fall into the category of metal/metal oxide nanoparticles. Magnetic imaging for biological applications is the trend of the day and particles that cater to these needs are iron oxide/silica, iron oxide/gold nanocomposite. The former is a metal oxide/inorganic nanocomposite while latter is a metal oxide/metal nanocomposite.

The gold coating on iron oxide makes it microchip compatible and the silica coat prevents agglomeration and also enables conjugation. Some other core/shell nanoparticles in this category include tin/tin oxide nanoparticles that are used in food processing and humidity sensor, gold/ palladium core/shell structure which are used as catalysts.

1.2. Semiconductor Nanoparticles

These types of nanoparticles have core made of semiconductor material, semiconductor alloy or metal oxide with shell made of semiconductor material, metal oxide or inorganic material like silica. These structures can be binary with a core and shell or a tertiary structure with a core and two shells. The most common binary structure that are well known by the name quantum dots are an alloy of group 3 and group 5 metals or group 4 and group 6 metals. CdSe/Cds, CdSe/ZnS, ZnSe/ZnS, CdTe/CdS are the nanoparticles used for fluorescent bioimaging. The shell thickness determines the emission range of these particles. They fall under the category of binary nanoparticles. The CdSe/CdTe core/shell nanoparticles are stable and have high conductivity. Addition of acceptor ions such as bipyridium to CdSe/Zns or similar nanoparticles can enhance the fluorescence properties. Mechanical properties can be improved by adding high performance polyamines and polypyrroles. The electronic properties can be enhanced and size tunability of quantum dot nanoparticles depends on the matrix material. Nanoquantum dots can be dispersed in Titania matrix and they have better luminescence than the quantum dots in sols. Quantum dots have magnetic property when the core shell structure has Co/Cd or Cd/Co. There are complex structures such ZnS, ZnxCd1-xS in PMMA matrix which show variation in luminescence on varying Zn/Cd ratio. There are also tertiary structures that have magnetic property such as iron oxide/ CdSe/ZnSe. These are bifunctional having both fluorescence and magnetic properties. Thus detailed *in vivo* imaging is possible.

1.3. Lanthanide Nanoparticles

These particles core which contains one or more lanthanide group elements surrounded by a shell made of inorganic material like silica or a lanthanide material. Rhabdophane lanthanide phosphate aqueous colloids are one of these categories that show a green luminescence. These are Ce; Tb doped core particles with an LnPO4-xH2O shell. These particles can be further coated with silica to enhance its luminescent properties. These find potential application in electronics and bioimaging. Other lanthanide particles include YF3/Silica, TiO2: Eu phosphors and the like. The YF3 particles show emission in the range of 400nm and TiO2/Eu phosphors show a red emission peak. The potential application of these particles is in the field of bioimaging.

2. Organic-Inorganic Hybrid Core/Shell Nanoparticles

2.1. Organic core and Inorganic Shell Nanoparticles

The core of these particles consists of organic compounds and can be polymers of organic compounds. The shell is inorganic and is a metal or silica or silicone. Structures that fall under the category of polymer/ metal are polyethylene/silver, polylactide/gold. These are used in joint replacements and due to their high resistance to corrosion and abrasion can be used to improve properties of other materials. Many of these nanoparticles manufactured fall under the category of polymer/silica and some of them are polycaprolactum/silica, polystyrenemethylmethacrylate/ silica.

2.2 Inorganic Core and Organic Shell Nanoparticles

The particles in this class of nanoparticles have a metal, metal oxide or silica core with a polymer of organic material or organic shell. Some of the particles in this category are SiO2/PAPBA(Poly(3-aminophenylboronic acid), Ag2S/ PVA(Polyvinylalcohol), CuS/PVA, Ag2S/PANI(Poly aniline), and TiO2/cellulose. The coat of PAPBA prevents agglomeration of particles and helps maintain size control.SiO2/PAPBA is used in optical devices, sensors and electrical devices. The PVA or PANI coat prevents oxidation of Ag2S /CuS thus improving stability. Ag2S /CuS are conducting polymers that are used in electrical industry. The addition of cellulose improves the pigment properties of TiO2.Some of these hybrids find application in dentistry as brace material and fillers.

2.3. Polymeric Core/Shell Nanoparticles

These are particles that have a polymeric core and a polymeric shell and are dispersed in a matrix which can be any material whose property is to be modified. One of the materials in this category is

Polymethylmethacrylate (PMMA) coated antimony trioxide compounded with Polyvinylchloride (PVC)/antimony trioxide composites. The interaction between PMMA and the PVC along with antimony trioxide enhance toughness and strength of PVC. Some of these particles improve the thermal sensitivity of materials. Polystyrene/ Poly (N-isopropylacrylamide) (P-NIPA) core/shell nanoparticles with silver embedded in the NIPA coat are found to enhance the catalytic activity of silver by improving sensitivity. In the field of electronics the sensitivity to voltage change is improved by using junctions such as (poly-(3,4 dicyanothiophene) PDCTh / (methoxy-5- (2′-ethyl-hexyloxy)-1, 4-phenylene vinylene) MEH-PPV. Phase segregated nanoparticles blend of donor (MEH-PPV) and Cyano substituted phenylene vinylene (CNPPV) show enhanced photo induced charge separation. Some of the other composite latexes that find application are PSt/PVAc, PSt/ Ppy and PSt/PBA where PSt is polystyrene. PVAc is polyvinyl acetate, Ppy is polypyrrole and PBA is poly butyl alcohol.

3. Biomedical Applications and Patents

Core/Shell Nanoparticles are finding wide spread applications in all fields. They are making their way into our day to day life. Things such as cabinet and car doors contain nanoparticles which improve their durability. At a large scale the industries that make most use of these materials are the chemical, electronics, biomedical, civil and mechanical Industries. They are used as catalysts, modifiers, fillers, thermal and mechanical property enhancers, sensor material due to high sensitivity to slight changes in parameter. The biomedical industry is the potential play field today; hence some of the possible application will be highlighted. Some of these applications are patented and some are still in the research phase. The diverse branches of the industry where nanoparticles are being recognized are bioimaging, drug delivery, biomarkers and transplants.

3.1 Bio-imaging

The imaging modalities in which core/shell nanoparticles are used are MRI and luminescence. Magnetic nanoparticles iron oxide or cobalt core particles are used to enhance MRI images by improving contrast. Core/shell nanoparticles can enter the cells and they have better spin-lattice relaxation time. Thus the contrast is better. These particles are found to be biocompatible, so they seem very promising. The luminescent particles are particles that fluoresce by absorbing light over in a wavelength range and emit in the visible or near-IR range. The nanoparticles with NIR region emission are still being worked on as the cells do not emit in the same range. Thus signal to noise ratio is high. Though they cannot be easily applied to in-depth live cell imaging, the application is not impossible. Similarly particles that emit in the UV-visible range sometimes have low signal to noise ratio. But the composition and particle size can be varied to get appropriate emission and detection is also a lot easier. The present trend are the up conversion nanoparticles which show anti-stokes emission and these are brighter than down conversion nanoparticles that show stokes emission. The ranges of nanoparticles that are luminescent are semiconductor nanoparticles, lanthanide based nanoparticles, and gold coated silica nanoparticles and the like. Multifunctional nanoparticles that have magnetic and luminescent properties can also be synthesized by choosing appropriate core and shell. This helps in obtaining clear 3D images.

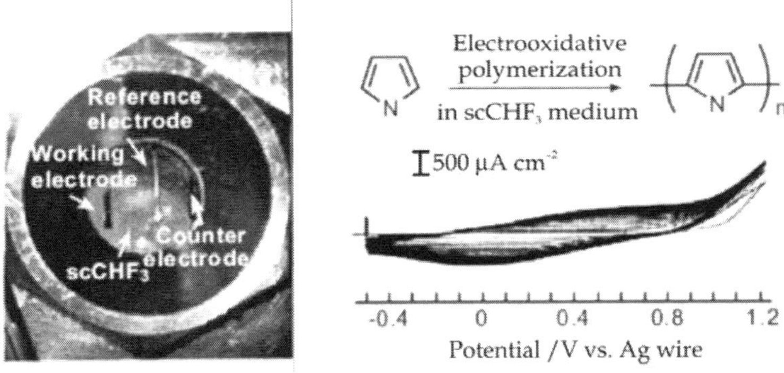

Fig.5.1 : Electrochemical Synthesis of Nanoparticles (Reprinted with permission from Atobe et al. copyright 2004. The Chemical Society of Japan)

3.2 Drug Delivery

Core/Shell structures serve a wide category of drug delivery application. The particles are biocompatible, have the ability to be conjugated to molecules without affecting the core and also can be used to encapsulate drugs. The material of choice decides the multifunctional nature of the particles. They can be structured so that they can be used for imaging and for drug delivery. Drug eluting stents made of nanocomposite material are being worked upon to reduce abrasion and also core/shell nanoparticles with antibacterial properties are also being worked upon. The latter seems promising for the manufacture of catheters which shall reduce infections that are caused by catheter contamination. The drugs can be targeted to specific locations by attaching biomolecules such as antibodies to the surface of nanoparticles. This is very useful when it comes to targeting tumor cells. Thermo sensitive and pH sensitive nanoparticles can be used for environment controlled delivery of drug from the particles. Bifunctional nanoparticles with a luminescent core and shell conjugated with a biomolecule/drug that can be used for targeting /drug delivery and imaging. Drug loaded contact lenses are an elucidation of how these materials can be used for encapsulating drugs.

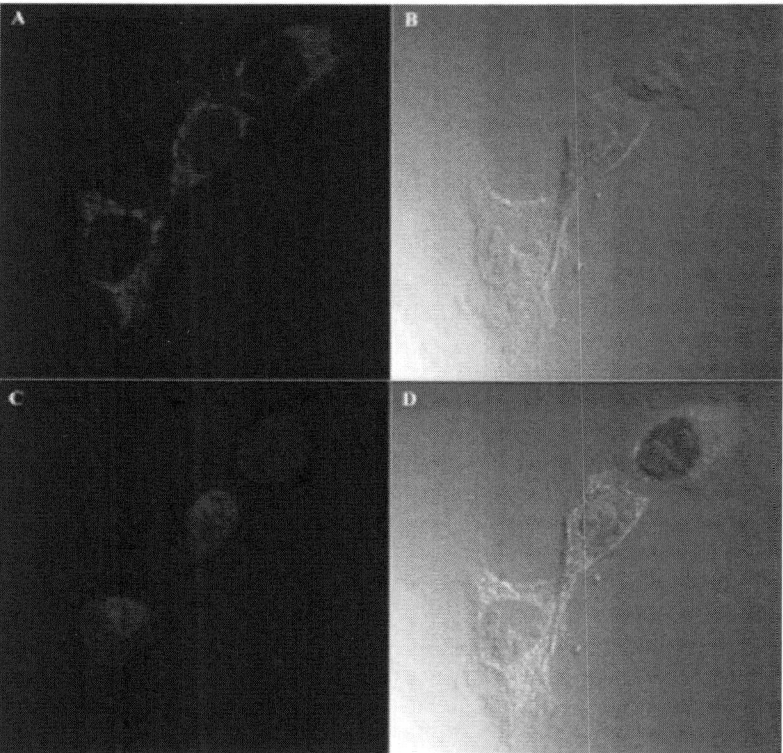

Fig. 5.2 : Uptake of Nanoparticles by 143B Osteosarcoma Cells. A) Detection of Texas Red Labeled Nanoparticles, B) Bright Field, C) Detection of DAPI and D) Overlap of these Three Images (Reprinted with permission from Azarmi et al. copyright 2006. Canadian Society for Pharmaceutical Sciences)

3.3 Cell Labeling

Nano biomarkers will be more efficient as they can operate at wide pH and temperature ranges. They are small and can penetrate through cells and even if they do not penetrate they can be targeted to specific cells by attaching suitable moieties to the shell surface. The wide color range that these particles display on varying shell thickness and their photo stability are also important characters that make them suitable for marking/cell labeling. Nanocomposite biomarkers are used for labeling cancer cells and tumor cells. There can be multifunctional particles that act not only as markers, but also can be used for cell separation, if the core of the particle has magnetic properties.

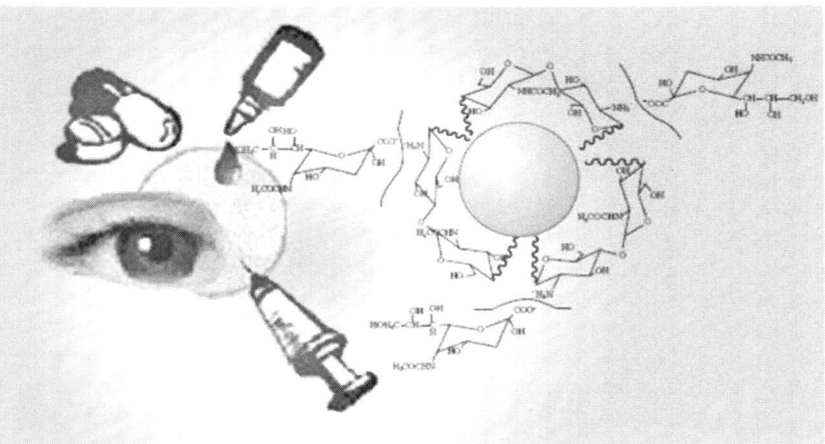

Fig. 5.3 : Nanoparticles in Ophthalmic Drug Delivery (Reprinted with permission from Barbu et al. Copyright 2006. Royal Society of Chemistry)

Replacements, Supports and Engineered Tissues

Polymeric core shell nanocomposite is the most commonly used transplant material. These can be polymer/ polymer or polymer/metal, either ways forming core/shell structures. They are used in dental braces and in joint replacements. Ultra high molecular weight polyethylene(UHMWPE)/silver is one such material that's used in joint replacement. The salient features that are considered in the choice of these materials for joints are abrasion resistance, high impact strength and resistance to corrosion. Biomimetic nanoparticles seem to be better materials for joint replacements as they closely resemble the biological structure and behave almost similarly.

But they are a bit expensive and are still at the research stage. Some nanoparticles are being worked on to be used as artificial conjunctiva. Tissue engineering is also looking up to nanoparticles for design of suitable scaffold for seeding and proliferation of the cells to develop artificial tissues. This shows the salient features of nanoparticles that makes it a preferred choice for different purposes with suitable modification. Materials used commonly are hydroxyapatite based, collagen based or ceramic based nanoparticles.

Miscellaneous Application

Nano-nanoparticles can be used for lipid and protein detection by modifying their surface with suitable charged or specific binding moieties. This is a boon because it reduces size of the testing kits and offers high sensitivity.

They are used in medical dressing for quicker wound healing. These have the property of absorbing fluids and can be medicated that can help in consistent wound healing. There are nanoparticle structures that are used by the nutraceutical industry. They are expected to deliver nutrients more efficiently than normal drug capsules. They are also used to design filter material for better and efficient separation of components. This application is not restricted only to the biological industry. Other application that's common to both biological and the chemical industry is that of catalytic activity enhancement. Enzymes immobilized on core/shell nanoparticles or catalysts encapsulated suitably into nanoparticles have higher activity than usual. Some find applications as biosensors.

PLGA-PEG Block Copolymers for Drug Formulations

Recently, biodegradable polymers, especially poly (lactic acid) (PLA), poly(glycolic acid) (PGA), and poly(lactic-co-glycolic acid) (PLGA), have been used significantly in pharmaceutical and biomedical applications. Poly (lactic acid), poly (glycolic acid), and poly (lactic-co-glycolic acid) have also been called polylactide, polyglycolide, and poly (lactide-co-glycolide), respectively, according to the nomenclature system based on the source of the polymer. Although these names were used in many references in the past, a recent trend is to follow the nomenclature system of the International Union of Pure and Applied Chemistry (IUPAC) that is based on the repeating unit structure. PLA, PGA, and PLGA can be degraded into non-toxic substances and removed from the human body. Accordingly, they have taken center stages in a variety of research efforts.

The biodegradable polyesters are all strongly hydrophobic, and this has caused some limitations in practical drug formulations. To add hydrophilic and other physico-chemical properties, poly (ethylene glycol) (PEG) has been incorporated into the biodegradable polyesters. PEG is a non-toxic, water-soluble polymer with proven biocompatibility. Block copolymers consisting of a hydrophobic polyester segment and a hydrophilic PEG segment have attracted large attention due to their biodegradability, biocompatibility, and tailor-made properties. A wide variety of drug formulations, such as micro/nano-particles, micelles, hydrogels, and injectable drug delivery systems have been developed using PLGA-PEG block copolymers. They have been extensively investigated for use in a wide range of applications, including implantable materials, drug delivery systems, and tissue engineering scaffolds. They are very useful materials for pharmaceutical and biomedical applications, and thus they have significant commercial potential. Demands for the block copolymers will continue to grow, and they are expected to build a great market as their precursors, such as PLA, PGA, and PLGA, did.

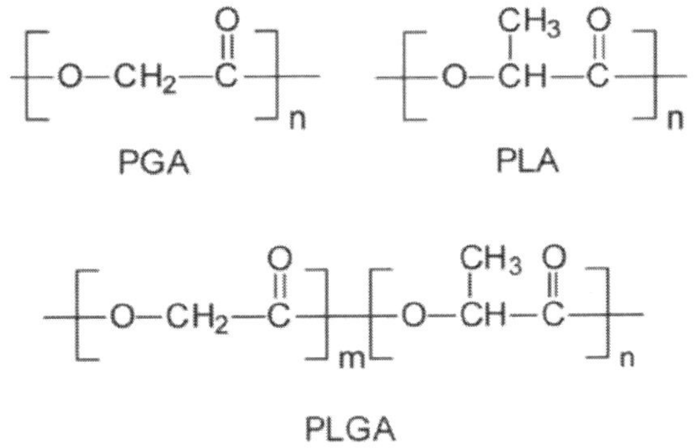

$$\left[-O-CH_2-\overset{\overset{O}{\|}}{C}-\right]_n \qquad \left[-O-\overset{\overset{CH_3}{|}}{CH}-\overset{\overset{O}{\|}}{C}-\right]_n$$

PGA PLA

$$\left[-O-CH_2-\overset{\overset{O}{\|}}{C}-\right]_m\left[-O-\overset{\overset{CH_3}{|}}{CH}-\overset{\overset{O}{\|}}{C}-\right]_n$$

PLGA

Fig. 5.4 : Chemical structure of poly (glycolic acid), poly (lactic acid) (PLA,) and poly (Lactic-co-glycoid acid (PLGA).

PLA, PGA & PLGA as Biomaterials

Biodegradable polymers have been extensively used in controlled drug delivery. They have the advantage of not requiring surgical removal after they serve their intended purposes. PGA, PLA, and especially their copolymers PLGA are the most commonly used family of biodegradable polymers. PGA was used as a biodegradable suture material in the 1970s, and it has led the largest volume production in the biomedical polymer markets, when its production was combined with those of PLA and PLGA. Since then, they have found a broad range of pharmaceutical and biomedical applications based on their unique properties, including versatile degradation kinetics, non-toxicity, and biocompatibility. The general properties and typical applications of PGA, PLA, and PLGA are summarized in Table 5.1.

Table 5.1 : Properties and application of PGA, PLA, and PLGA [S,22]

Polymer	Crystallinity	T0(°C)	Degradation Rate°	Typical application
PGA	Highly crystalline (Tm = 225~230°C)	35~40	2-3 months	Suture, Soft anaplerosis
PLA (f. form)	Semicrystalline (T_m = 173~178°C)	60~65	> 2 years	Fracture fixation, Lagament augmentation
PLA (D, L form)	Amorphous	55~60	12-16 months	Drug delivery system
PLGA	Amorphous	45~55	1-6 months[b]	Suture, Fracture Fixation, Oral implant, Drug delivery microsphare

a. Rate depends on molecular weights of the polymers
b. Rate may change according to the ration of LA and GA

PGA is a highly crystalline polymer and the most hydrophilic among them. It has a very high melting point (224°C to 226°C), and the degradation rate of PGA is much higher than that of PLA. Random PLGA copolymers with different ratios of lactide (LA) and glycolide (GA) exhibit different degradation rates, and thus can be tailor-made for specific applications requiring specific degradation kinetics ranging from weeks to months. They are generally more amorphous than their homo-polymers and become most susceptible to hydrolysis when the two monomer contents are the same.

Properties of PLGA-PEG Block Copolymers

One typical characteristic of PLGA degradation is autocatalysis by which heterogeneous bulk degradation is observed with a decrease in pH. Carboxylic end groups of degraded products, oligomeric PLGA, can accelerate the degradation and decrease the local pH in PLGA formulations. This locally acidified environment is known to be a main reason for protein inactivation and often requires incorporation of antacids, such as Mg $(OH)_2$, into the polymers for protein stabilization by neutralization. In addition, other limitations, such as hydrophobicity, brittleness, and toxicity, have been reported with PLGA formulations. Block copolymers containing hydrophilic PEG segments have attracted considerable attention as an alternative approach for overcoming such undesirable effects and improving the properties in applications of PLGA as drug delivery vehicles. Through various synthetic processes, diverse block copolymers with a wide range of molecular weights, chemical structure, and hydrophilic/hydrophobic block ratios have been prepared and used in controlled drug delivery.

The chemical composition and molecular weight of block copolymers determines their water-solubility and degradation kinetics. Polymers with low molecular weight or composed of shorter hydrophobic blocks are soluble in water, whereas high molecular weight polymers and polymers with longer hydrophobic blocks are not soluble but swell in water. In general, the degradation time will be shorter for low molecular weight polymers, more hydrophilic polymers, more amorphous polymers, and copolymers with higher content of glycolide. Therefore, at identical conditions, low molecular weight copolymers of lactide and glycolide will degrade relatively rapidly, whereas the high molecular weight homopolymers, PLA, and PGA will degrade much more slowly.

PLGA-PEG Block Copolymers in drug formulations

Polymeric nanoparticles are nanoparticles which are prepared from polymers. The drug is dissolved, entrapped, encapsulated or attached to nanoparticles and depending upon the method of preparation, nanoparticles, nanospheres or nanocapsules can be obtained. Nanocapsules are vesicular systems in which the drug is confined to a cavity surrounded by a polymer membrane, while nanospheres are matrix systems in which the drug is physically and uniformly dispersed.

In recent years, biodegradable polymeric nanoparticles have attracted considerable attention as potential drug delivery devices in view of their applications in drug targeting to particular organs/tissues, as carriers of DNA in gene therapy, and in their ability to deliver proteins, peptides and genes through oral route of administration.

In spite of development of various synthetic and semi synthetic polymers, natural polymers still enjoy their popularity in drug delivery, some of them are: gums (Eg. Acacia, Guar, etc.,) chitosan, gelatin, sodium alginate, and albumin. A range of materials have been employed to delivery of bioactive agents. A polymer used in controlled drug delivery formulation, must be chemically inert, non-toxic and free of leachable impurities. It must also have an appropriate physical structure, with minimal undesired aging, and be readily processable. Some of the polymeric materials are: cellulosics., poly(2-hydroxy ethyl methacrylate), poly(N-vinyl pyrrolidone), poly(methyl methacrylate), poly(vinyl alcohol), poly(acrylic acid), polyacrylamide, poly(ethylene-co-vinyl acetate), poly(ethylene glycol), poly(methacrylic acid). However, in recent years additional polymers are designed primarily for medical applications and have entered the arena of controlled release of bioactive agents. Many of these materials are designed to degrade within the body, most popular ones are: polylactides (PLA), polyglycolides (PGA), poly(lactide-co-glycolides) (PLGA), polyanhydrides, polyorthoesters, polycyanoacrylates, polycaprolactone. Originally, polylactides and polyglycolides were used as absorbable suture material. The main advantage of these degradable polymers is that they are broken down into biologically acceptable molecules that are metabolized and removed from the body via normal metabolic pathways. However, biodegradable materials do produce degradation by-products that must be tolerated with little or no adverse reactions within the biological environment.

Abraxane is the first polymeric nanoparticle based product from American Pharmaceutical Partners Inc., and American Bioscience, Inc. (ABI). It was approved in year 2005 and consists of albumin-bound paclitaxel nanoparticles. This product is free of toxic solvents like cremophor-EL, which is used until now to solubulize paclitaxel in order to administer it intravenously to the patient. Cremophor-EL is known to cause life-threatening allergic reactions. Success of Abraxane shows that nanotechnology can bring many exciting products which can overcome many hurdle of formulation.

Drug Formulations

Various types of drug formulations, such as nano/micro-particles, hydrogels, micelles, and injectable delivery systems have been developed using PLGA-PEG block copolymers to deliver hydrophobic drugs as well as hydrophilic peptide and protein drugs. The ability of the polymers to entrap drugs and subsequently release them at a controlled rate has been used to develop various drug formulations.

Nano/Microparticles

ABA triblock copolymers are more hydrophilic than PLA or PLGA itself, and are considered more suitable for development of delivery systems for hydrophilic macromolecular drugs, such as peptides, proteins, and oligo/polynucleotides. Micro- and nano-particles prepared from AB diblock and ABA triblock copolymers are extensively investigated for protein drug delivery. Block copolymer nano/microparticles are generally prepared by water/oil/water (W/O/W) double emulsion method, as shown in Figure 4. The W1 phase, an aqueous phase containing protein drugs, is dispersed into the oil phase consisting of polymer dissolved in organic solvent (eg, dichloromethane) using a high-speed homogenizer. The primary water-in-oil (W/O) emulsion is then dispersed to an aqueous solution containing a polymeric surfactant, eg, poly(vinyl alcohol) (PVA), and further homogenized to produce a W/O/W emulsion. After stirring for several hours, the nano/microparticles are collected by filtration. They show quite different release patterns when compared with PLGA. Kissel at al. reported that ABA triblock copolymers showed a continuous and molecular mass-dependent release while the release from PLGA was biphasic and almost independent of the molecular mass of entrapped substances.

Polymeric Micelles

Amphiphilic PLGA-PEG block copolymers form micelles composed of a hydrophobic PLGA core and hydrophilic PEG shell in water, as sho vn in Figure 5. Hydrophobic blocks are segregated from the aqueous exterior to form an inner core surrounded by a palisade of hydrophilic segments. Block copolymer micelles are water-soluble, biocompatible nanocontainers in the size of 10~100 nm with proven efficacy of delivering hydrophobic drugs. These micelle-forming block copolymers can provide high concentration of hydrophobic drugs with increased drug stability in an aqueous milieu above the solubility limit of the drug. The ability of polymeric micelles to target certain cells (eg, cancer cells) can also lower the required dosage. The size and morphology of block copolymer micelles can be easily changed by adjusting the chemical composition, total molecular weight, and ratio of the block lengths. Various hydrophobic drugs, including paclitaxel, have been incorporated into the hydrophobic inner core of micelles.

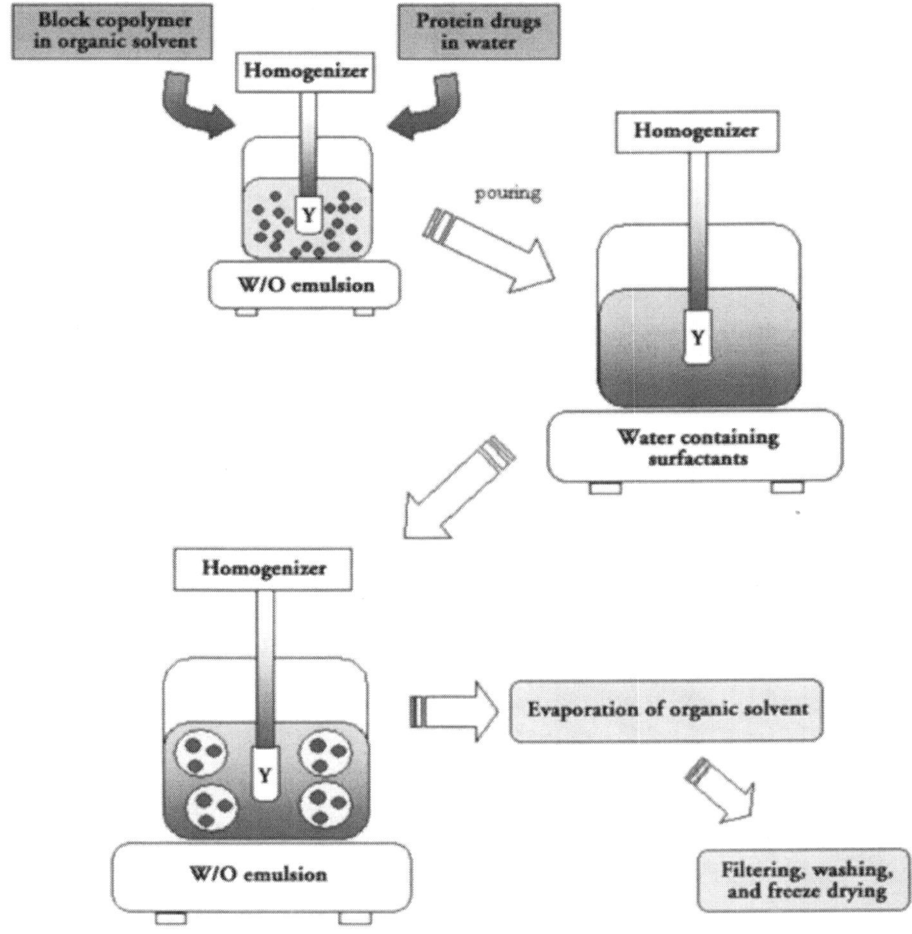

Fig. 5.5 : Schematic representation of a general procedure for PLGA-PEG block co-polymer nano/microparticles.

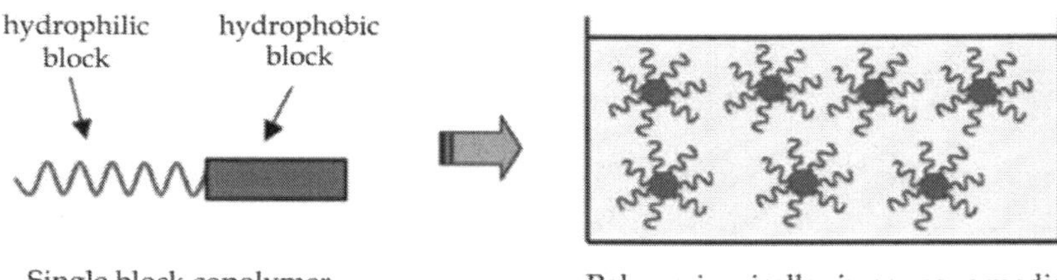

Fig. 5.6 : Self assembly of amphiphilic block copolymers to a micelle structure in aqueous solution.

Hydrogels

When the block copolymers have high molecular weights and high PLGA contents, they become water-insoluble but can swell in water. Block copolymers consisting of hydrophilic and hydrophobic blocks are able to form physical crosslinking in an aqueous environment through hydrophobic interaction; crystalline microdomains or chain entanglement. Physical associations of hydrophobic domains maintain swollen soft domains together and keep the polymer network stable in water. Although physical associations are reversible and weaker than chemical cross linking, they allow solvent casting and thermal processing, and the resulting polymer gels often possess elastic or viscoelastic properties. These biodegradable, physical hydrogels may offer an alternative material of choice in designing drug delivery systems as well as other biomedical applications.

Injectable Drug Delivery Systems

Aqueous solutions of low molecular weight B-A-B type triblock copolymers are well known to have thermo-reversible sol-gel transitions, forming in situ hydrogels without harmful organic solvents or any chemical reactions. Figure 6 schematically represents the thermo-reversible sol-gel transition of the triblock copolymers. The system can be loaded with drugs in aqueous phase at low temperature (below critical gelation temperature) where it forms a sol. Just following subcutaneous injection, the elevated temperature to 37°C (above critical gelation temperature) makes the injected sol to a gel that can act as a sustained-release matrix for the loaded drugs. Thermo-reversible hydrogels have recently attracted large attention due to the simplicity of drug formulation by solution mixing, biocompatibility with biological systems, and convenient administration. Pharmaceutical and biomedical applications of the block copolymers include solubilization of low molecular weight hydrophobic drugs, controlled release of labile biomacromolecules (eg, proteins and genes), cell immobilization, and tissue engineering.

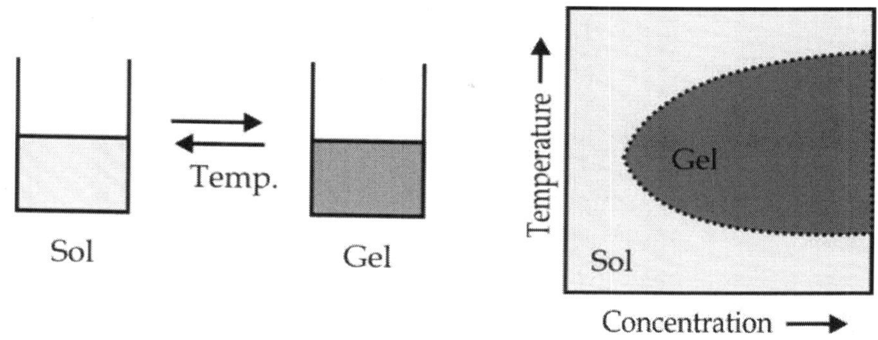

Lack of Commercial Sources for PLGA-PEF Block Copolymers

Biodegradable polymers have been playing a key role in various pharmaceutical and biomedical research and product development. Materials that can degrade and disappear from the body are desirable for a number of applications, including orthopedics, tissue engineering, and controlled drug delivery systems. For this reason, a number of commercial sources supply PLA, PGA, and PLGA throughout the world. However, there are no commercial sources supplying block copolymers consisting of hydrophobic blocks of PLA, PGA, or PLGA and hydrophilic PEG blocks. These block copolymers have unique properties that homopolymers cannot provide, and are ideal for formulating controlled drug delivery systems and for making scaffolds for tissue engineering. Yet their commercial introduction has not been made to date. Demands for these block copolymers are expected to grow as the pharmaceutical and biomedical fields continue to grow.

Commercial Production of Plga-peg Copolymers

During the past few decades, significant advances have been made in polymeric drug delivery technology. Drug delivery in the future will need more sophisticated and diverse formulations for existing drugs as well as new drugs, such as protein and peptide drugs. Biodegradable polymers have an important role in the development of controlled-release formulations. PLGA-PEG block copolymers have a number of unique and useful properties that are ideal for controlled drug delivery. The properties of sol-gel transition at well-defined temperatures and biodegradability make block copolymers an ideal delivery vehicle for various drugs, including protein drugs. Numerous new protein drugs are expected in the near future as a result of an increased understanding of genomics and proteomics. Thus, demands for the PLGA-PEG block copolymers are expected to grow exponentially for both research and product development in the coming years. To meet the current needs and prepare for the future demands, Akina, Inc., started producing biodegradable PLGA-PEG block copolymers for those who wish to use them in their research and development of drug delivery systems (www.akinainc.com/polycelle).

Synthetic methods for the block copolymers are well established, and Akina produces block copolymers consisting of glycolide, lactide, and PEG. These three main building blocks can be assembled into a wide variety of different combinations to synthesize block copolymers with diverse, but tailor-made, properties. Properties, such as degradation rate, hydrophilicity, cystallinity, and solubility, can be easily custom-made for specific applications.

❑❑❑

Chapter-6

Fullerenes and Carbon Nano Materials

A fullerene is any molecule composed entirely of carbon, in the form of a hollow sphere, ellipsoid, or tube. Spherical fullerenes are also called buckyballs, and cylindrical ones are called carbon nanotubes or buckytubes. Fullerenes are similar in structure to graphite, which is composed of stacked graphene sheets of linked hexagonal rings; but they may also contain pentagonal (or sometimes heptagonal) rings.

The first fullerene to be discovered, and the family's namesake, was buckminsterfullerene C_{60}, made in 1985 by Robert Curl, Harold Kroto and Richard Smalley. The name was homage to Richard Buckminster Fuller, whose geodesic domes it resembles. Fullerenes have since been found to occur (if rarely) in nature.

The discovery of fullerenes greatly expanded the number of known carbon allotropes, which until recently were limited to graphite, diamond, and amorphous carbon such as soot and charcoal. Buckyballs and buckytubes have been the subject of intense research, both for their unique chemistry and for their technological applications, especially in materials science, electronics, and nanotechnology.

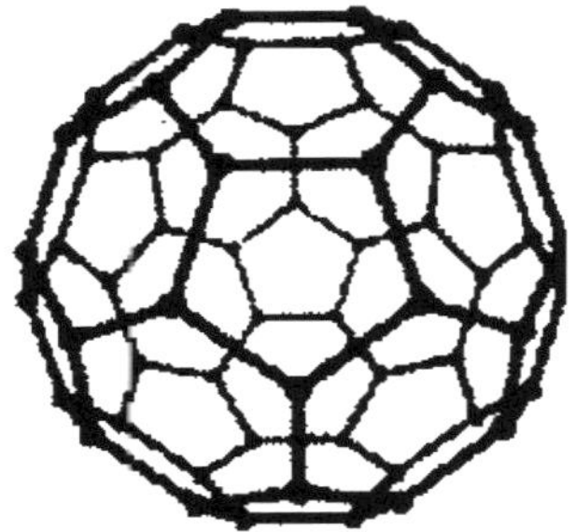

Fig. 6.1 : Buckminsterfullerene C_{60}

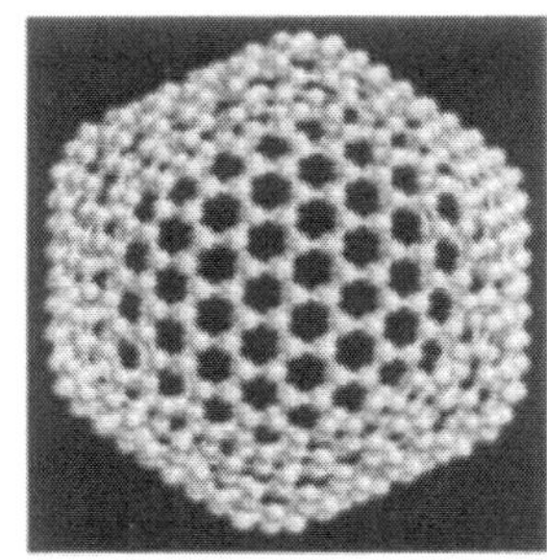

Fig. 6.2 : The Icosahedral Fullerene C_{540}

Prediction and discovery

The existence of C_{60} was predicted by Eiji Osawa of Toyohashi University of Technology in a Japanese magazine in 1970. He noticed that the structure of a corannulene molecule was a subset of a soccer-ball shape, and he made the hypothesis that a full ball shape could also exist. His idea was reported in Japanese magazines, but did not reach Europe or America.

With mass spectrometry, discrete peaks were observed corresponding to molecules with the exact mass of sixty or seventy or more carbon atoms. In 1985, Harold Kroto (then of the University of Sussex), James R. Heath, Sean O'Brien, Robert Curl and Richard Smalley, from Rice University, discovered C_{60}, and shortly thereafter came to discover the fullerenes. Kroto, Curl, and Smalley were awarded the 1996 Nobel Prize in Chemistry for their roles in the discovery of this class of compounds. C_{60} and other fullerenes were later noticed occurring outside the laboratory (e.g., in normal candle soot). By 1991, it was relatively easy to produce gram-sized samples of fullerene powder using the techniques of Donald Huffman and Wolfgang Krätschmer. Fullerene purification remains a challenge to chemists and to a large extent determines fullerene prices. So-called endohedral fullerenes have ions or small molecules incorporated inside the cage atoms. Fullerene is an unusual reactant in many organic reactions such as the Bingel reaction discovered in 1993. The first nanotubes were obtained in 1991.

Minute quantities of the fullerenes, in the form of C_{60}, C_{70}, C_{76}, and C_{84} molecules, are produced in nature, hidden in soot and formed by lightning discharges in the atmosphere. Recently, fullerenes were found in a family of minerals known as Shungites in Karelia, Russia.

Naming

Buckminsterfullerene (C_{60}) was named after Richard Buckminster Fuller, a noted architectural modeler who popularized the geodesic dome. Since buckminsterfullerenes have a similar shape to that sort of dome, the name was thought to be appropriate. As the discovery of the fullerene family came *after* buckminsterfullerene, the shortened name 'fullerene' was used to refer to the family of fullerenes.

Variations

Since the discovery of fullerenes in 1985, structural variations on fullerenes have evolved well beyond the individual clusters themselves. Examples include:

- *Buckyball clusters*: smallest member is C_{20} (unsaturated version of dodecahedrane) and the most common is C_{60};
- *Nanotubes:* hollow tubes of very small dimensions, having single or multiple walls; potential applications in electronics industry;
- *Megatubes:* larger in diameter than nanotubes and prepared with walls of different thickness; potentially used for the transport of a variety of molecules of different sizes;
- *Polymers:* chain, two-dimensional and three-dimensional polymers are formed under high pressure high temperature conditions
- *Nano"onions"*: spherical particles based on multiple carbon layers surrounding a buckyball core; proposed for lubricants;
- *Linked "ball-and-chain" dimers:* two buckyballs linked by a carbon chain;
- Fullerene rings

"Buckyball"

Buckminsterfullerene (IUPAC name (C_{60}-I_h) fullerene) is the smallest fullerene molecule in which no two pentagons share an edge (which can be destabilizing, as in pentalene). It is also the most common in terms of natural occurrence, as it can often be found in soot. The structure of C_{60} is a truncated (T = 3) icosahedron, which resembles a soccer ball of the type made of twenty hexagons and twelve pentagons, with a carbon atom at the vertices of each polygon and a bond along each polygon edge. The van der Waals diameter of a C_{60} molecule is about 1 nanometer (nm). The nucleus to nucleus diameter of a C_{60}

molecule is about 0.7 nm. The C_{60} molecule has two bond lengths. The 6:6 ring bonds (between two hexagons) can be considered "double bonds" and are shorter than the 6:5 bonds (between a hexagon and a pentagon). Its average bond length is 1.4 angstroms.

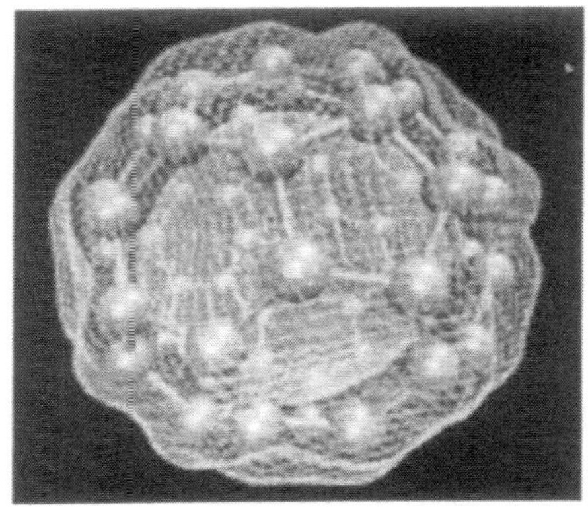

Fig. 6.3 : C_{60} with isosurface of ground state electron density as calcuelatd with DFT

Fig. 6.4 : An association football is a model of the Buckminsterfullerene C_{60}

Boron buckyball

A new type of buckyball utilizing boron atoms instead of the usual carbon has been predicted and described by researchers at Rice University. The B-80 structure, with each atom forming 5 or 6 bonds, is predicted to be more stable than the C-60 buckyball. One reason for this is that the B-80 is actually more like the original geodesic dome structure popularized by Buckminster Fuller which utilizes triangles rather than hexagons. However, this work has been subject to much criticism by quantum chemists as it was concluded that the predicted Ih symmetric structure was vibrationally unstable and the resulting cage undergoes a spontaneous symmetry break yielding a puckered cage with rare Th symmetry (symmetry of a volleyball). The number of six atom rings in this molecule is 20 and number of five member rings is 12. There is an additional atom in the center of each six member ring, bonded to each atom surrounding it.

Variations of buckyballs

Another fairly common buckminsterfullerene is C70, but fullerenes with 72, 76, and 84 and even up to 100 carbon atoms are commonly obtained.

In mathematical terms, the structure of a fullerene is a trivalent convex polyhedron with pentagonal and hexagonal faces. In graph theory, the term fullerene refers to any 3-regular, planar graph with all faces of size 5 or 6 (including the external face). It follows from Euler's polyhedron formula, $|V| - |E| + |F| = 2$, (where $|V|$, $|E|$, $|F|$ indicate the number of vertices, edges, and faces), that there are exactly 12 pentagons in a fullerene and $|V|/2-10$ hexagons.

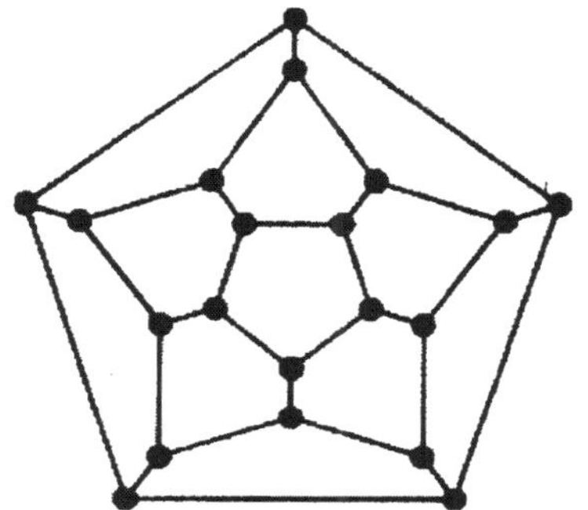

Fig. 6.5 : 20- fullerene (dodecahedral graph)

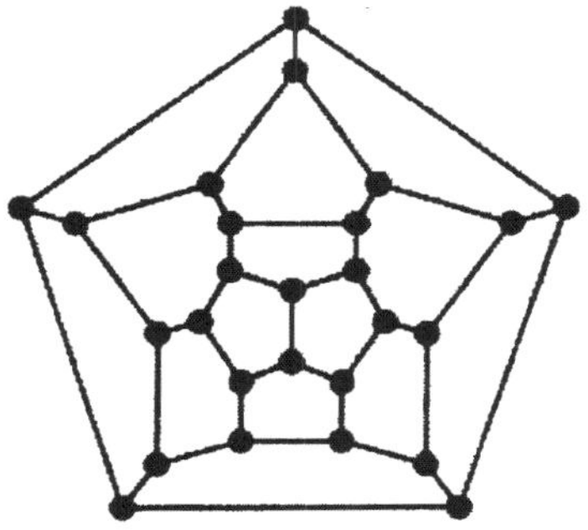

Fig. 6.6 : 26-fullerene graph

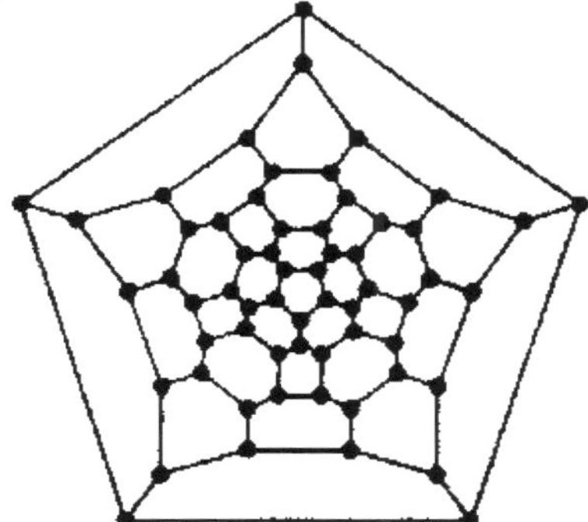

Fig. 6.7 : 60-fullerene (truncated icosahedral graph)

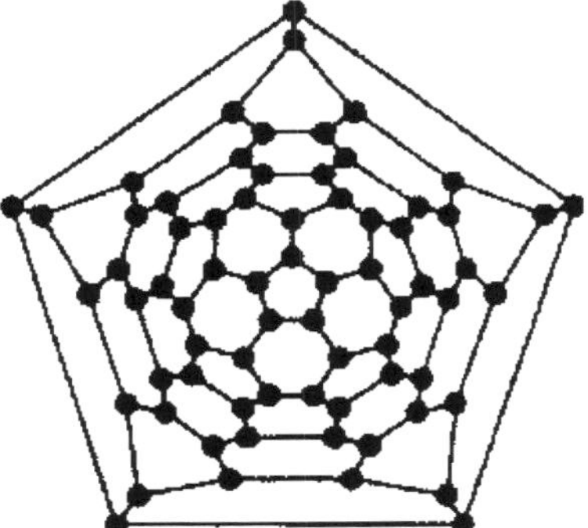

Fig. 6.8 : 70-fullerene graph

illustrate the growth, there are 214,127,713 non-isomorphic fullerenes C200, 15,655,672 of which have no adjacent pentagons.

Trimetasphere carbon nanomaterials were discovered by researchers at Virginia Tech and licensed exclusively to Luna Innovations. This class of novel molecules comprises 80 carbon atoms (C80) forming a sphere which encloses a complex of three metal atoms and one nitrogen atom. These fullerenes encapsulate metals which puts them in the subset referred to as metallofullerenes. Trimetaspheres have the potential for use in diagnostics (as safe imaging agents), therapeutics and in organic solar cells.

Carbon nanotubes

Nanotubes are cylindrical fullerenes. These tubes of carbon are usually only a few nanometres wide, but they can range from less than a micrometer to several millimeters in length. They often have closed ends, but can be open-ended as well. There are also cases in which the tube reduces in diameter before closing off. Their unique molecular structure results in extraordinary macroscopic properties, including high tensile strength, high electrical conductivity, high ductility, high resistance to heat, and relative chemical inactivity (as it is cylindrical and "planar" - that is, it has no "exposed" atoms that can be easily displaced). One proposed use of carbon nanotubes is in paper batteries, developed in 2007 by researchers at Rensselaer Polytechnic Institute. Another proposed use in the field of space technologies and science fiction is to produce high-tensile carbon cables required by a space elevator.

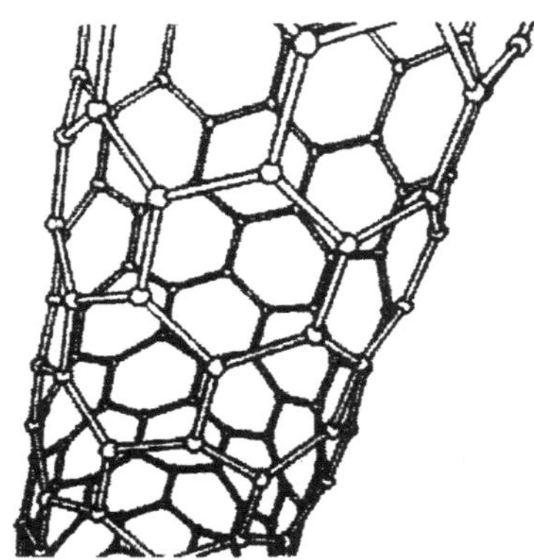

Fig. 6.9 : This model of a rotating Carbon nanotube shows its 3D structure.

Methodology

A common method used to produce fullerenes is to send a large current between two nearby graphite electrodes in an inert atmosphere. The resulting carbon plasma arc between the electrodes cools into sooty residue from which many fullerenes can be isolated. For the past decade, the chemical and physical properties of fullerenes have been a hot topic in the field of research and development, and are likely to continue to be for a long time. Popular Science has published articles about the possible uses of fullerenes in armor. In April 2003, fullerenes were under study for potential medicinal use: binding specific antibiotics to the structure to target resistant bacteria and even target certain cancer cells such as melanoma. The October 2005 issue of Chemistry and Biology contains an article describing the use of fullerenes as light-activated antimicrobial agents. In the field of nanotechnology, heat resistance and superconductivity are some of the more heavily studied properties.

Chemistry

Fullerenes are stable, but not totally unreactive. The sp2-hybridized carbon atoms, which are at their energy minimum in planar graphite, must be bent to form the closed sphere or tube, which produces angle strain. The characteristic reaction of fullerenes is electrophilic addition at 6, 6-double bonds, which reduces angle strain by changing sp2-hybridized carbons into sp3-hybridized ones. The change in hybridized orbitals causes the bond angles to decrease from about 120 degrees in the sp2 orbitals to about 109.5 degrees in the sp3 orbitals. This decrease in bond angles allows for the bonds to bend less when closing the sphere or tube, and thus, the molecule becomes more stable.

Other atoms can be trapped inside fullerenes to form inclusion compounds known as endohedral fullerenes. An unusual example is the egg shaped fullerene Tb3N@C84, which violates the isolated pentagon rule. Recent evidence for a meteor impact at the end of the Permian period was found by analyzing noble gases so preserved. Metallofullerene-based inoculates using the rhonditic steel process are beginning production as one of the first commercially-viable uses of buckyballs.

Solubility

Fullerenes are sparingly soluble in many solvents. Common solvents for the fullerenes include aromatics, such as toluene, and others like carbon disulfide. Solutions of pure buckminsterfullerene have a deep purple color. Solutions of C70 are a reddish brown. The higher fullerenes C76 to C84 have a variety of colors. C76 has two optical forms, while other higher fullerenes have several structural isomers. Fullerenes are the only known allotrope of carbon that can be dissolved in common solvents at room temperature.

Some fullerene structures are not soluble because they have a small band gap between the ground and excited states. These include the small fullerenes C28, C36 and C50. The C72 structure is also in this class, but the endohedral version with a trapped lanthanide-group atom is soluble due to the interaction of the metal atom and the electronic states of the fullerene. Researchers had originally been puzzled by C72 being absent in fullerene plasma-generated soot extract, but found in endohedral samples. Small band gap fullerenes are highly reactive and bind to other fullerenes or to soot particles.

Solvents that are able to dissolve buckminsterfullerene (C60) are listed below in order from highest solubility. The value in parentheses is the approximate saturated concentration.

1. 1-chloronaphthalene (51 mg/mL)
2. 1-methylnaphthalene (33 mg/mL)
3. 1,2-dichlorobenzene (24 mg/mL)
4. 1,2,4-trimethylbenzene (18 mg/mL)
5. tetrahydronaphthalene (16 mg/mL)
6. carbon disulfide (8 mg/mL)
7. 1,2,3-tribromopropane (8 mg/mL)
8. bromoform (5 mg/mL)
9. cumene (4 mg/mL)
10. toluene (3 mg/mL)
11. benzene (1.5 mg/mL)
12. cyclohexane (1.2 mg/mL)
13. carbon tetrachloride (0.4 mg/mL)
14. chloroform (0.25 mg/mL)
15. n-hexane (0.046 mg/mL)
16. tetrahydrofuran (0.006 mg/mL)
17. acetonitrile (0.004 mg/mL)

18. methanol (0.000 04 mg/mL)
19. water (1.3x10?11 mg/mL)

Solubility of C_{60} in some solvents shows unusual behaviour due to existence of solvate phases (analogues of crystallohydrates). For example, solubility of C_{60} in benzene solution shows maximum at about 313 K. Crystallization from benzene solution at temperatures below maximum results in formation of triclinic solid solvate with four benzene molecules $C_{60}°4C_6H_6$ which is rather unstable in air. Out of solution, this structure decomposes into usual fcc C_{60} in few minutes' time. At temperatures above solubility maximum the solvate is not stable even when immersed in saturated solution and melts with formation of fcc C_{60}. Crystallization at temperatures above the solubility maximum results in formation of pure fcc C_{60}. Large millimetre size crystals of C_{60} and C_{70} can be grown from solution both for solvates and for pure fullerenes.

Quantum mechanics

In 1999, researchers from the University of Vienna demonstrated that wave-particle duality applied to molecules such as fullerene. One of the co-authors of this research, Julian Voss-Andreae, has since created several sculptures symbolizing wave-particle duality in fullerenes. Science writer Marcus Chown stated on the CBC radio show 2006 that scientists are trying to make buckyballs exhibit the quantum behavior of existing in two places at once (quantum superposition).

Safety and toxicity

When considering toxicological data, care must be taken to distinguish as necessary between what are normally referred to as fullerenes: (C_{60}, C_{70},...); fullerene derivatives: C_{60} or other fullerenes with covalently bonded chemical groups; fullerene complexes (e.g., water-solubilized with surfactants, such as C_{60}-PVP; host-guest complexes, such as with cyclodextrin;), where the fullerene is physically bound to another molecule; C_{60} nanoparticles, which are extended solid-phase aggregates of C_{60} crystallites; and nanotubes, which are generally much larger (in terms of molecular weight and size) compounds, and are different in shape to the spheroidal fullerenes C_{60} and C_{70}, as well as having different chemical and physical properties.

The above different compounds span the range from insoluble materials in either hydrophilic or lipophilic media, to hydrophilic, lipophilic, or even amphiphilic compounds, and with other varying physical and chemical properties. Therefore any broad generalization extrapolating for example results from C60 to nanotubes or vice versa is not possible, though technically all are fullerenes, as the term is defined as a close-caged all-carbon molecule. Any extrapolation of results from one compound to other compounds must take into account considerations based on a Quantitative Structural Analysis Relationship Study (QSARS), which mostly depends on how close the compounds under consideration are in physical and chemical properties.

In 1996 and 1997, Moussa et al. studied the in vivo toxicity of C60 after intra-peritoneal administration of large doses. No evidence of toxicity was found and the mice tolerated a dose of 5,000 mg/kg of body weight. Mori et al. (2006) could not find toxicity in rodents for C60 and C70 mixtures after oral administration of a dose of 2,000 mg/kg body weight and did not observe evidence of genotoxic or mutagenic potential in vitro. Other studies could not establish the toxicity of fullerenes: on the contrary, the work of Gharbi et al. (2005) suggested that aqueous C60 suspensions failing to produce acute or subacute toxicity in rodents could also protect their livers in a dose-dependent manner against free-radical damage.

A comprehensive and recent review on fullerene toxicity is given by Kolosnjaj et al. (2007). These authors review the works on fullerene toxicity beginning in the early 1990s to present, and conclude that very little evidence gathered since the discovery of fullerenes indicate that C60 is toxic.

With reference to nanotubes, a recent study of Poland et al. (2008) on carbon nanotubes introduced into the abdominal cavity of mice led the authors to suggest comparisons to "asbestos-like pathogenicity". It should be noted that this was not an inhalation study, though there have been several performed in the past, therefore it is premature to conclude that nanotubes should be considered to have a toxicological profile similar to asbestos. Conversely, and perhaps illustrative of how the various classes of compounds which fall under the general term fullerene cover a wide range of properties, Sayes, et al.(2007) found that in *vivo* inhalation of $C_{60}(OH)_{24}$ and nano-C_{60} in rats gave no effect, whereas in comparison quartz particles produced an inflammatory response under the same conditions. As stated above, nanotubes are quite different in chemical and physical properties to C60, i.e., molecular weight, shape, size, physical properties (such as solubility) all are very different, so from a toxicological standpoint, different results for C60 and nanotubes are not suggestive of any discrepancy in the findings.

Superconductivity

After the synthesis of macroscopic amounts of fullerenes, their physical properties was investigated. Very soon Haddon et al. found that intercalation of alkali-metal atoms in solid C60 leads to metallic behavior. In 1991, it was revealed that potassium-doped C60 becomes superconducting at 18K. This was the highest transition temperature for a molecular superconductor. Since then, superconductivity has been reported in fullerene doped with various other alkali metals. It has been shown that the superconducting transition temperature in alkaline-metal-doped fullerene increases with the unit-cell volume. As cesium forms the largest alkali ion, cesium-doped fullerene is an important material in this family. Recently, superconductivity at 38K has been reported in bulk Cs3C60, but only under applied pressure. The highest superconducting transition temperature of 33 K at ambient pressure is reported for Cs2RbC60.

Chirality

Few fullerenes (e.g. C76, C78, C80, and C84) are inherently chiral because they are D2-symmetric and have been successfully resolved. Research efforts are ongoing to develop specific sensors for their enantiomers.

Fullerite (solid state)

Fullerites are the solid-state manifestation of fullerenes and related compounds and materials.

Fig. 6.10 : The C_{60} fullerene in crystalline form

Ultrahard fullerite, Buckyball

"Ultrahard fullerite" is a coined term frequently used to describe material produced by high-pressure high-temperature (HPHT) processing of fullerite. Such treatment converts fullerite into a nanocrystalline form of diamond which exhibits remarkable mechanical properties.

Carbon nano materials

Allotropes of carbon

Eight allotropes of carbon are there

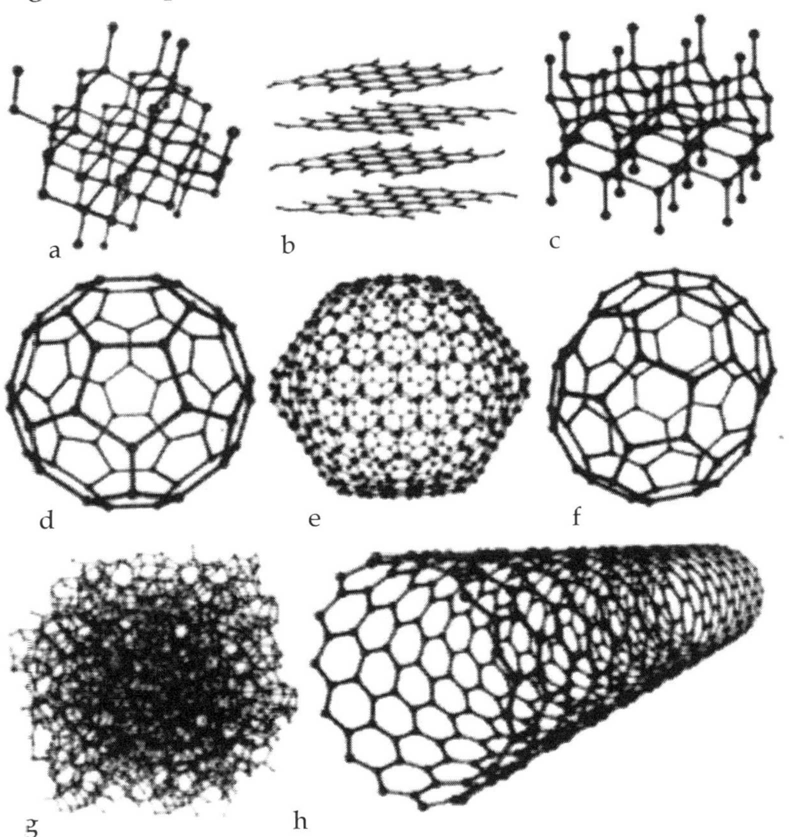

Fig. 6.11 : Eight allotropes of carbon: a) Diamond, b) Graphite, c) Lonsdaleite, d) C_{60} (Buckminsterfullerene or buckyball), e) C_{540}, f) C_{70}, g) Amorphous carbon, and h) single-walled carbon nanotube or buckytube.

Diamond

Diamond is one of the best known allotropes of carbon, whose hardness and high dispersion of light make it useful for industrial applications and jewellery. Diamond is the hardest known natural mineral, which makes it an excellent abrasive and makes it hold polish and luster extremely well. The market for industrial-grade diamonds operates much differently from its gem-grade counterpart. Industrial diamonds are valued mostly for their hardness and heat conductivity, making many of the gemological characteristics of diamond, including clarity and color, mostly irrelevant. This helps explain why 80% of mined diamonds, unsuitable for use as gemstones and known as bort, are destined for industrial use. In addition to mined diamonds, synthetic diamonds found industrial applications almost immediately after their invention in the 1950s.

The dominant industrial use of diamond is in cutting, drilling (drill bits), grinding (diamond edged cutters), and polishing. Most uses of diamonds in these technologies do not require large diamonds. With the continuing advances being made in the production of synthetic diamond, future applications are beginning to become feasible.

Each carbon atom in a diamond is covalently bonded to four other carbons in a tetrahedron. These tetrahedrons together form a 3-dimensional network of six-membered carbon rings (similar to cyclohexane), in the chair conformation, allowing for zero bond angle strain. This stable network of covalent bonds and hexagonal rings, is the reason that diamond is so incredibly strong.

Graphite

Graphite (named by Abraham Gottlob Werner in 1789, from the Greek anUoaei (*graphein*, "to draw/write", for its use in pencils) is one of the most common allotropes of carbon. Unlike diamond, graphite is an electrical conductor, and can be used, for instance, as the material in the electrodes of an electrical arc lamp. Graphite holds the distinction of being the most stable form of carbon under standard conditions.

Graphite is able to conduct electricity, due to delocalization of the pi bond electrons above and below the planes of the carbon atoms. These electrons are free to move, so are able to conduct electricity. However, the electricity is only conducted along the plane of the layers. Graphite powder is used as a dry lubricant. Although it might be thought that this industrially important property is due entirely to the loose interlamellar coupling between sheets in the structure, in fact in a vacuum environment (such as in technologies for use in space), graphite was found to be a very poor lubricant. This fact led to the discovery that graphite's lubricity is due to adsorbed air and water between the layers, unlike other layered dry lubricants such as molybdenum disulfide. Recent studies suggest that an effect called superlubricity can also account for this effect. When a large number of crystallographic defects bind these planes together, graphite loses its lubrication properties and becomes what is known as pyrolytic carbon, a useful material in blood-contacting implants such as prosthetic heart valves. In its pure glassy (isotropic) synthetic forms, pyrolytic graphite and carbon fiber graphite is an extremely strong, heat-resistant (to 3000 °C) material.

Graphene

A single layer of graphite, once thought to be impossible, is called graphene and has extraordinary electrical, thermal, and physical properties. It can be produced by epitaxy (vapor deposition) on an insulating surface or by mechanical exfoliation (repeated peeling). Its applications may include replacing silicon in high-performance electronic devices.

Amorphous carbon

Amorphous carbon is the name used for carbon that does not have any crystalline structure. As with all glassy materials, some short-range order can be observed, but there is no long-range pattern of atomic positions. While entirely amorphous carbon can be produced, most amorphous carbon actually contains microscopic crystals of graphite-like, or even diamond-like carbon.

Coal and soot or carbon black are informally called amorphous carbon. However, they are products of pyrolysis (the process of decomposing a substance by the action of heat), which does not produce true amorphous carbon under normal conditions. The coal industry divides coal up into various grades depending on the amount of carbon present in the sample compared to the amount of impurities. The highest grade, anthracite, is about 90% carbon and 10% other elements. Bituminous coal is about 75-90% carbon, and lignite is the name for coal that is around 55% carbon.

Buckminsterfullerenes

The buckminsterfullerenes, or usually just fullerenes for short, were discovered in 1985 by a team of scientists from Rice University and the University of Sussex, three of whom were awarded the 1996 Nobel Prize in Chemistry. They are named for the resemblance of their alliotropic structure to the geodesic structures devised by the scientist and architect Richard Buckminster "Bucky" Fuller. Fullerenes are molecules of varying sizes composed entirely of carbon, which take the form of a hollow sphere, ellipsoid, or tube.

Buckyballs

Spherical fullerenes are popularly known as buckyballs, and have the formula C60.

Carbon nanotubes

Carbon nanotubes, also called buckytubes, are cylindrical carbon molecules with novel properties that make them potentially useful in a wide variety of applications (e.g., nano-electronics, optics, materials applications, etc.). They exhibit extraordinary strength, unique electrical properties, and are efficient conductors of heat. Inorganic nanotubes have also been synthesized. A nanotube is a member of the fullerene structural family, which also includes buckyballs. Whereas buckyballs are spherical in shape, a nanotube is cylindrical, with at least one end typically capped with a hemisphere of the buckyball structure. Their name is derived from their size, since the diameter of a nanotube is on the order of a few nanometers (approximately 50,000 times smaller than the width of a human hair), while they can be up to several centimeters in length. There are two main types of nanotubes: single-walled nanotubes (SWNTs) and multi-walled nanotubes (MWNTs).

Carbon nanobuds

Carbon NanoBuds are a newly discovered allotrope of carbon in which fullerene like "buds" are covalently attached to the outer sidewalls of the carbon nanotubes. This hybrid material has useful properties of both fullerenes and carbon nanotubes. In particular, they have been found to be exceptionally good field emitters.

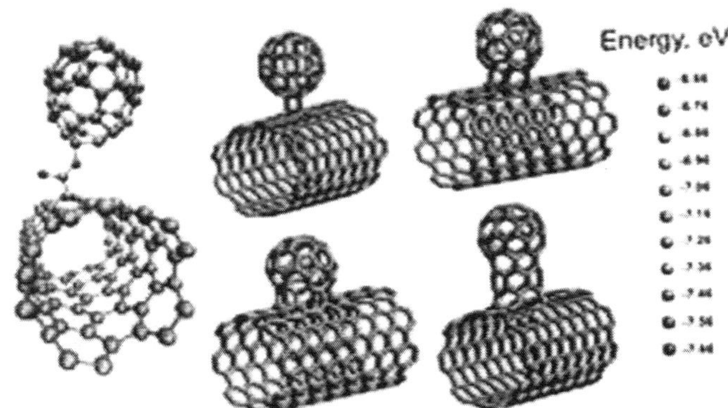

Fig. 6.12 : Computer models of stable NanoBud structures

Carbon nanofoam

Carbon nanofoam is the fifth known allotrope of carbon discovered in 1997 by Andrei V. Rode and co-workers at the Australian National University in Canberra. It consists of a low-density cluster-assembly of carbon atoms strung together in a loose three-dimensional web. Each cluster is about 6 nanometers wide and consists of about 4000 carbon atoms linked in graphite-like sheets that are given negative curvature by the inclusion of heptagons among the regular hexagonal pattern. This is the opposite of what happens in the case of buckminsterfullerenes, in which carbon sheets are given positive curvature by the inclusion of pentagons.

The large-scale structure of carbon nanofoam is similar to that of an aerogel, but with 1% of the density of previously produced carbon aerogels - only a few times the density of air at sea level. Unlike carbon aerogels, carbon nanofoam is a poor electrical conductor.

Nanotube membrane

Nanotube membrane is either a single, open-ended nanotube or a film composed of open-ended nanotubes that are oriented perpendicularly to the surface of an impermeable film matrix like the cells of a honeycomb. 'Impermeable' is essential here to distinguish nanotube membrane with traditional, well known porous membranes. Fluids and gas molecules may pass through the membrane en masse.

Transport of polystyrene particles (60 and 100 nm diameter) through single-tube membranes (150 nm) was reported back in 2000.

Soon after, ensemble membranes consisting of multi-wall carbon nanotubes, were fabricated and studied. It was shown that water can pass through the graphitic nanotube cores of the membrane at speeds several magnitudes greater than classical fluid dynamics would predict, both for multiwall tubes (inner diameter 7 nm) and double-wall tubes (inner diameter <2 nm). It was further demonstrated that the flow of water through carbon nanotube membranes (without filler matrix, thus flow on the outside surface of CNTs) can be controlled through the application of electrical current. Among many potential uses that nanotube membranes might one day be employed is the desalination of water.

Buckypaper

Buckypaper is a thin sheet made from an aggregate of carbon nanotubes. The nanotubes are approximately 50,000 times thinner than a human hair. Originally, it was fabricated as a way to handle carbon nanotubes, but it is also being studied and developed into applications by several research groups, showing promise as a building material for aerospace vehicles, body armor and next-generation electronics and displays.

Fig. 6.13 : Buckypaper made of carbon nanotubes

Synthesis

The generally accepted methods of making CNT films involves the use of non-ionic surfactants, such as Triton X-100 and Sodium lauryl sulfate, which improves their dispersibility in aqueous solution. These suspensions can then be membrane filtered under positive or negative pressure to yield uniform films. The Van der Waals force's interaction between the nanotube surface and the surfactant can often be mechanically strong and quite stable and therefore there are no assurances that all the surfactant is removed from the CNT film after formation. Washing with methanol, an effective solvent in the removal of Triton X, was found to cause cracking and deformation of the film. It has also been found that Triton X can lead to cell lysis and in turn tissue inflammatory responses even at low concentrations.

In order to avoid adverse side-effects from the possible presence of surfactants, an alternative casting process can be used involving a frit compression method that did not require the use of surfactants or surface modification. The dimensions can be controlled through the size of the syringe housing and the through the mass of carbon nanotubes added. Their thicknesses are typically much larger than surfactant-cast buckypaper and have been synthesized from 120 ?m up to 650 ?m; whilst no nomenclature system exists to govern thicknesses for samples to be classified as paper, samples with thicknesses greater than 500 ?m are referred to as buckydiscs. The frit compression method allows rapid casting of buckypaper and buckydiscs with recovery of the casting solvent and control over the 2D and 3D geometry.

Aligned multi-walled carbon nanotube (MWCNT) growth has been used in CNT film synthesis through the domino effect. In this process, "forests" of MWCNTs are pushed flat in a single direction, compressing their vertical orientation into the horizontal plane, which results in the formation of high-purity buckypaper with no further purification or treatment required. By comparison, when a buckypaper sample was formed from the 1 ton compression of chemical vapor deposition (CVD) generated MWCNT powder, any application of a solvent led to the immediate swelling of the film till it reverted into particulate matter. It appears that for the CNT powder used, compression alone was insufficient to generate robust buckypaper and highlights that the aligned growth methodology generates in-situ tube-tube interactions not found in CVD CNT powder and are preserved through to the domino pushing formation of buckypaper.

Properties

Buckypaper is one tenth the weight yet potentially 500 times stronger than steel when its sheets are stacked to form a composite. It could disperse heat like brass or steel and it could conduct electricity like copper or silicon.

Applications

Among the possible uses for buckypaper that are being researched:

- Fire protection: Covering any material with a thin layer of buckypaper significantly improves its fire resistance due to the efficient reflection of heat by the dense, compact layer of carbon nanotubes or carbon fibers.
- If exposed to an electric charge, buckypaper could be used to illuminate computer and television screens. It could be more energy-efficient, lighter, and could allow for a more uniform level of brightness than current cathode ray tube (CRT) and liquid crystal display (LCD) technology.
- Since individual carbon nanotubes are one of the most thermally conductive materials known, buckypaper lends itself to the development of heat sinks that would allow computers and other electronic equipment to disperse heat more efficiently than is currently possible. This, in turn, could lead to even greater advances in electronic miniaturization.
- Because individual carbon nanotubes have an unusually high current-carrying capacity, a buckypaper film could be applied to the exteriors of airplanes. Lightning strikes then could flow around the plane and dissipate without causing damage.
- Films also could protect electronic circuits and devices within airplanes from electromagnetic interference, which can damage equipment and alter settings. Similarly, such films could allow military aircraft to shield their electromagnetic "signatures", which can be detected via radar.
- Buckypaper could act as a filter membrane to trap microparticles in air or fluid. Because the nanotubes in buckypaper are insoluble and can be functionalized with a variety of functional groups, they can selectively remove compounds or can act as a sensor.
- Produced in high enough quantities and at an economically viable price, buckypaper composites could serve as an effective armor plating.
- Buckypaper can be used to grow biological tissue, such as nerve cells. Buckypaper can be electrified or functionalized to encourage growth of specific types of cells.

- The Poisson's ratio for carbon nanotube buckypaper can be controlled and has exhibited auxetic behavior, capable of use as artificial muscles.

❑❑❑

Chapter-7

Misnomers and Misconception of Nanotechnology

There are many different points of view about the nanotechnology. These differences start with the definition of nanotechnology. Some define it as any activity that involves manipulating materials between one nanometer and 100 nanometers. However the original definition of nanotechnology involved building machines at the molecular scale and involves the manipulation of materials on an atomic (about two-tenths of a nanometer) scale.

The debate continues with varying opinions about exactly what nanotechnology can achieve. Some researchers believe nanotechnology can be used to significantly extend the human lifespan or produce replicator-like devices that can create almost anything from simple raw materials. Others see nanotechnology only as a tool to help us do what we do now, but faster or better.

The third major area of debate concerns the timeframe of nanotechnology-related advances. Will nanotechnology have a significant impact on our day-to-day lives in a decade or two, or will many of these promised advances take considerably longer to become realities?

Finally, all the opinions about what nanotechnology can help us achieve echo with ethical challenges. If nanotechnology helps us to increase our lifespans or produce manufactured goods from inexpensive raw materials, what is the moral imperative about making such technology available to all? Is there sufficient understanding or regulation of nanotech based materials to minimize possible harm to us or our environment? Only time will tell how nanotechnology will affect our lives.

A common misconception about nanotech is that it is a single technology. Unlike biotechnology (which focuses on genes and DNA) or information technology (which focuses on microchips and software), nanotechnology encompasses a collection of methods and tools for dealing with all matter at the nano scale. It is best thought of as a new approach to building things. Working at the nano scale allows us to manufacture with unparalleled precision and efficiency. Rather than mining tons of ore at a great cost to the environment to find a handful of diamonds, nanotechnologists can start with carbon and build a flawless diamond one atom at a time. Because they are so precise, nanotech processes waste less material, consume less energy and produce better results.

A Crucial Misconception

1. We will be able to print out toys, parts, furniture, designs and more from the net using three-dimensional printers. Today 3D printers cost $20,000 and can only print prototypes. But maybe beyond 2013 you'll actually be able to print a pair of sneakers in your size. . .

   There will be no more lies. You'll still be able to have secrets but only if you can keep them off the net. Privacy will be available but only to those who can afford to pay for it. For most people, privacy will end in 2013, or a little beyond that. ..

   Quantum computing will shatter current encryption techniques, jeopardising anything that relies on encryption, such as credit card transactions, and requiring new approaches to information security. Using an electron spin as opposed to an electron charge in quantum computing would mean all the cryptography we have today would have to be rethought. . .

2. From 2025 to 2050 and beyond, nanotechnology will give us the capability to create anything, molecule by molecule, atom by atom. The technology is at very early stages but it is definitely going to happen. For example, it will be possible to create a piece of wood. But self-replication is crucial, otherwise the technology won't scale up. What are needed are nano parts that self-assemble — that way you can mass-produce anything.

This is a common misconception that still persists although it has been obsolete for a decade and half.

❑❑❑

Chapter-8

Methods of Nanoparticles Synthesis

Atoms and molecules are the essential building blocks of every object. The manner in which things are constructed with these basic units is vitally important to understand their properties and their reciprocal interactions. An efficient control of the synthetic pathways is essential during the preparation of nanobuilding blocks with different sizes and shapes that can lead to the creation of new devices and technologies with improved performances. Traditionally nanoparticles were produced only by physical and chemical methods. Some of the commonly used physical and chemical methods are ion sputtering, solvothermal synthesis, reduction and sol gel technique

Basically there are two approaches for nanoparticle synthesis namely the Bottom up approach and the Top down approach. In the Top down approach, scientists try to formulate nanoparticles using larger ones to direct their assembly. The Bottom up approach is a process that builds towards larger and more complex systems by starting at the molecular level and maintaining precise control of molecular structure.

These two different methods highlight the organization level of nanosystems as the crossing point hanging between the worlds of molecular objects and bulk materials.

The top-down approach has been advanced by Richard Feynman in his often-cited 1959 lecture stating that "there is plenty of room at the bottom" and it is ideal for obtaining structures with long-range order and for making connections with macroscopic world. Conversely, the bottom-up approach was pioneered by Jean-Marie Lehn (revealing that "there is plenty of room at the top") and it is best suited for assembly and establishing short-range order at the nanoscale. The integration of the two techniques is expected to provide, at least in principle, the widest combination of tools for nanofabrication.

Top-down Approach

The top-down approach is based on miniaturizing techniques, such as machining, templating or lithographic techniques. Top-down methods usually start from patterns generated at larger scale (generally at microscale) and then they are reduced to nanoscale. A key advantage of the top-down approach is that the parts are both patterned and built in place, so that no further assembly steps are needed.

By means of electronic, ionic or X-ray lithography, a monolith can be cut step by step in order to generate a *quantum well* (a bidimensional structure, with two finite dimensions) at first, then a *quantum wire* (a monodimensional structure, one finite dimension), and finally a *quantum dot* (a zero-dimensional structure, all the dimensions being in nanoscale). Current short-wavelength optical lithography methods can reach dimensions not less than 100 nanometers (the traditional threshold definition of the nanoscale). Extreme ultraviolet and X-ray sources are being developed to allow lithographic printing techniques to reach dimensions from 10 to 100 nanometers, but the principal limits are still due to the difficulty of beam focalization. Likewise, scanning beam techniques such as electron-beam lithography provide patterns down to about 20 nanometers and still-smaller features are obtained by using scanning probes to deposit or remove thin layers.

The general procedure of mechanical printing techniques consist on making a master "stamp" by a high-resolution lithographic technique, as described above, and then applying this stamp, or subsequent copies of it, to a surface to mould the pattern. The last step is to remove the thin layer of the masking material under the stamped regions. These nanoscale printing techniques offer several advantages due to the possibility to use a wide variety of materials with curved surfaces.

As a general drawback, these techniques are not cheap and require a complex manufacturing. In addition, top-down methodologies: (1), even if they work well at the microscale, they collide with some difficulties at nanoscale dimensions and (2) they usually lead to the formation of bidimensional structures and hence, as they are carried out by the addition or subtraction of patterned layers, they cannot easily give rise to the production of arbitrary three-dimensional objects.

It is worth underlining that the development in the top-down methodology was mainly driven from the traditional disciplines of materials engineering and physics, whereas the role of inorganic chemists has been minor in the exploitation of these techniques.

Bottom-up Approach

Bottom-up, or self-assembly, approaches to nanofabrication involve gradual additions of atoms or groups of atoms. This technique uses chemical or physical forces operating at the nanoscale to assemble basic

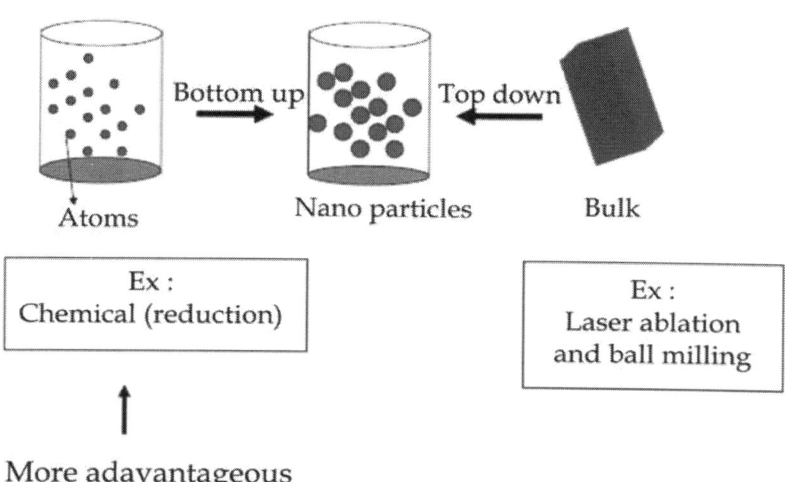

Fig. 8.1 : Bottom-up approach. The precursors (left) can react and assemble to form nanosystems with a large variety of shapes and sizes (right), depending on the reaction conditions.

units into larger structures. The chemical growth of nanometer-sized materials often implies colloidal or supramolecular systems and it frequently passes through phase transformations, such as vapor deposition on surfaces or precipitation of a solid phase from solution. Inspiration for bottom-up approaches comes from biological systems, where nature has employed chemical forces to create essentially all the structures needed by life. Researchers try to mimic nature's ability to produce small clusters of specific atoms, which can then self-assemble into more-elaborated structures.

In order to reach the desired shape and dimension of the new nanosystem, the nucleation and growth of the material have to be directed and controlled. This synthesis approach is of sure the most stimulating for the chemist. Indeed, bottom-up approaches, starting from single atoms and molecules have more affinity with chemistry and molecular biology, as much of chemistry already implicates the control of nanodimensional objects or the self-assembly of molecules into larger structures.

As main advantages, bottom-up techniques display a wide variety of preparation methods, they allow a good control onto scale dimension, even from atomic or molecular level, and they are not as expensive as top-down approaches. Nevertheless, the advent of pre-programmed self-assembling of arbitrarily large systems, with complexity comparable to that found in natural systems, is still a challenge.

In the following chapters, the attention will be focused only onto inorganic nanosized structures and the nanosystems will be defined and classified according to the hierarchical order of dimensionality: zero-dimensional systems, including spherical, pseudo-spherical or point-like objects; one-dimensional systems, in which the extension over one dimension is predominant over the other two, such as nanorods, nanowires, nanofibers, nanotubes; two-dimensional systems, such as flat or membrane-like materials, nanosheets and nanoscale discs; three-dimensional systems, both crystalline and amorphous nanostructures as well as porous and non-porous materials, which exhibit nanometric features, even if they extend over the three dimensions.

Classification of preparation methods

In terms of phase of medium for preparation
-Gas phase/ Liquid phase/ Aerosol phase/ Solid phase

In terms of method of "monomer" preparation
-Physical/ Chemical

Physical methods

- ❑ Electric arc discharge
- ❑ Flame pyrolysis
- ❑ Ball milling
- ❑ Laser ablation

Chemical methods

- ❑ Sol-gel methods
- ❑ Wet chemical co-precipitation
- ❑ Microencapsulation
- ❑ Hydrothermal methods
- ❑ Microwave synthesis
- ❑ Sonochemistry

Commen Physical and chemical methods of nanoparticle synthesis

Some of the commonly used physical and chemical methods include:

a) Sol-gel technique, which is a wet chemical technique used for the fabrication of metal oxides from a chemical solution which acts as a precursor for integrated network (gel) of discrete particles or polymers. The precursor sol can be either deposited on the substrate to form a film, cast into a suitable container with desired shape or used to synthesize powders.

b) Solvothermal synthesis, which is a versatile low temperature route in which polar solvents under pressure and at temperatures above their boiling points are used.

   Under solvothermal conditions, the solubility of reactants increases significantly, enabling reaction to take place at lower temperature.

c) Chemical reduction, which is the reduction of an ionic salt in an appropriate medium in the presence of surfactant using reducing agents. Some of the commonly used reducing agents are sodium borohydride, hydrazine hydrate and sodium citrate.

d) Laser ablation, which is the process of removing material from a solid surface by irradiating with a laser beam. At low laser flux, the material is heated by absorbed laser energy and evaporates or sublimates. At higher flux, the material is converted to plasma. The depth over which laser energy is absorbed and the amount of material removed by single laser pulse depends on the material's optical properties and the laser wavelength. Carbon nanotubes can be produced by this method.

e) Inert gas condensation, where different metals are evaporated in separate crucibles inside an ultra high vacuum chamber filled with helium or argon gas at typical pressure of few 100 pascals. As a result of inter atomic collisions with gas atoms in chamber, the evaporated metal atoms lose their kinetic energy and condense in the form of small crystals which accumulate on liquid nitrogen filled cold finger. E.g. gold nanoparticles have been synthesized from gold wires.

There are four fundamental routes to making nano materials.

Form in place

These techniques incorporate lithography, vacuum coating and spray coating.

Mechanical

This is a 'top-down' method that reduces the size of particles by attrition, for example, ball milling or planetary grinding.

Gas phase synthesis

These include plasma vaporization, chemical vapor synthesis and laser ablation.

Wet chemistry

This is the range of techniques that are most applicable for characterization by light scattering techniques. These are fundamentally 'bottom-up' techniques, i.e. they start with ions or molecules and build these up into larger structures.These nanoparticle manufacturing techniques historically come under the title of 'colloid chemistry', and involve classical 'sol-gel' processes, or other aggregation processes.These wet chemistry techniques currently offer the best quality nanoparticles from a number of points of view.

- They produce nanoparticles that are already in the form of dispersion, hence high inter-particle forces can be designed in to prevent agglomeration.
- The formation of aggregates can be reduced or eliminated.
- The nanoparticles can be made to be very monodisperse, i.e. all the same size to within small tolerances.
- The chemical composition, and morphology can be closely controlled. This is especially important for research purposes where the quality of the material must be very high to ensure repeatable and meaningful results.

One of the most recent developments is the production in liquid carbon dioxide. This offers the promise of the controlled conditions of the 'bottom-up' wet chemistry approach, as well as the benefit of being able to remove the dispersant by simply reducing the pressure of the reaction container. This technique is currently used for removing the caffeine from tea and coffee, so the mechanics of handling the materials are well understood.

❑❑❑

Chapter-9

Physical Methods of Nanoparticles Synthesis

It is classified as bottom-up manufacturing which involves building up of the atom or molecular constituents as against the top method which involves making smaller and smaller structures through etching from the bulk material as exemplified by the semiconductor industry.

I Physical Vapour Deposition Techniques (PVD)

Thermal Evaporation

Electron-Beam

RF induction

Resistive

Sputtering

Focussed ion beam

Radio frequency

Magnetron

Pulsed Laser Deposition

Ii Chemical Vapour Deposition (CVD)

Thermal CVD

Low pressure CVD (LPCVD)

Plasma enhanced CVD (PECVD)

Metal-organic CVD (MOCVD)

Molecular beam Epitaxy (MBE)

Atomic layer deposition (ALD)

III Solution Based Techniques (SBC)

Hydrothermal

Thermal Evaporation Process

The source materials are heated at its vapour point or above. Due to evaporation, the vaporized source gets deposited on the cooler portion of the substrate.

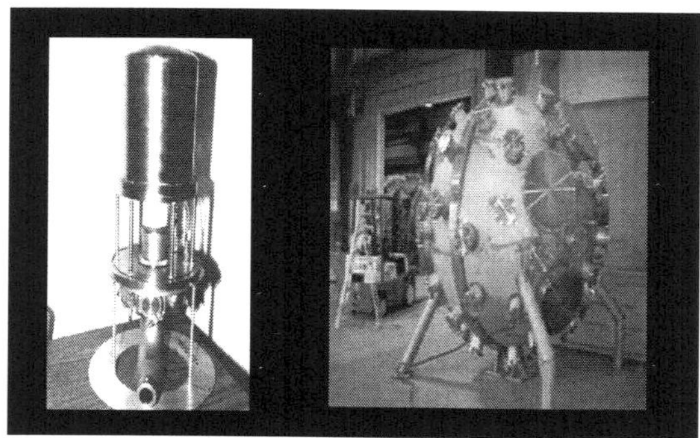

Fig. 9.1 : Thermal Evaporation Chamber

Sputtering Process

Sputtering is the removal of atomized material from a solid due to energetic bombardment of its surface layers by ions or neutral particles.

Focused Ion Beam

FIB uses a focused beam of gallium ions source to build a gallium liquid metal ion source (LMIS). These ions are then accelerated to an energy of 5-50 keV (kilo electron volts), and then focused onto the sample by electrostatic lenses. The FIB is destructive to the specimen. When the high-energy gallium ions strike the sample, they will sputter atoms from the surface. Gallium atoms will also be implanted into the top few nanometers of the surface, and the surface will be made amorphous.

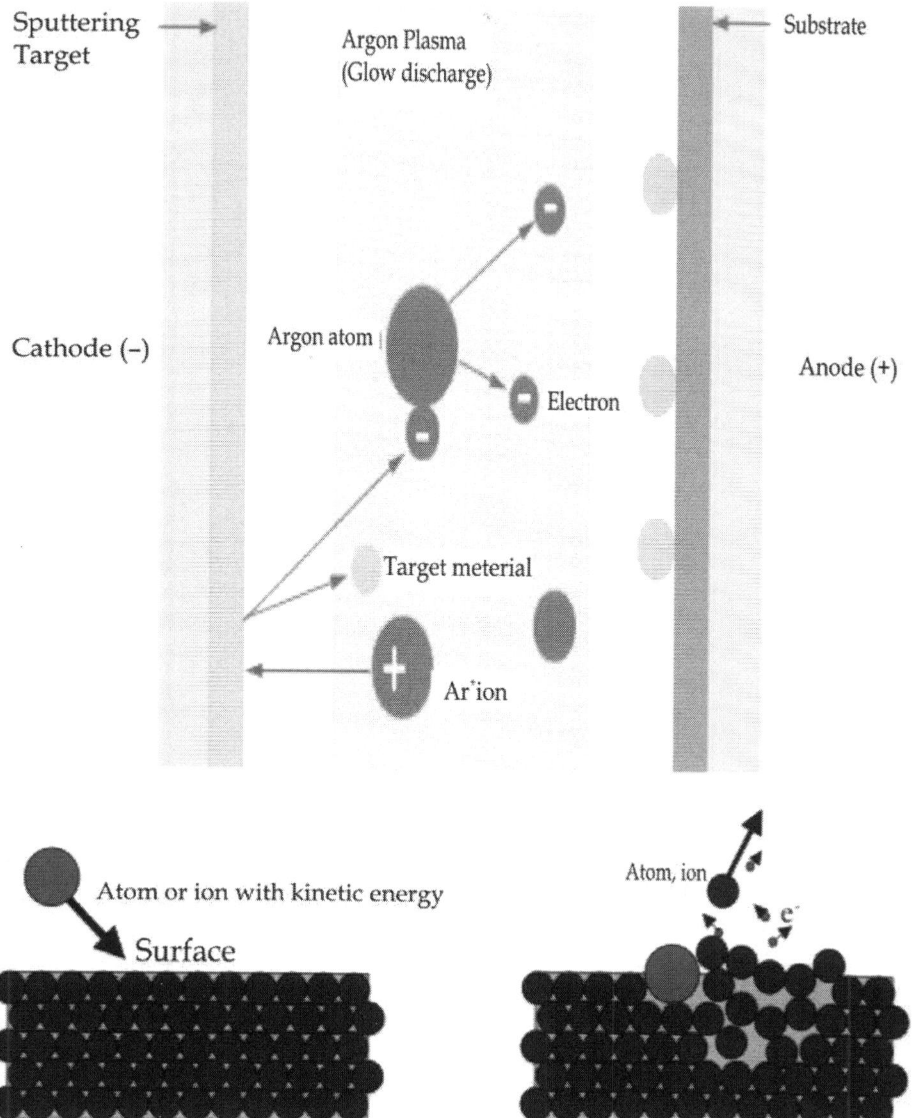

Fig. 9.2 : Sputtering Process

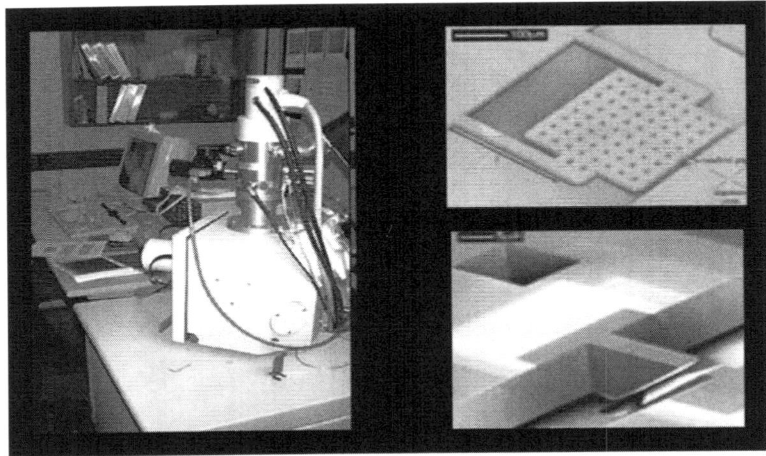

Fig. 9.3 : Focused ion beam (FIB)

Radio Frequency Sputtering

The substrate is normally placed in a low-pressure chamber between two electrodes.The electrodes are driven by an RF power source, which generate a plasma and ionization of the gas (e.g., argon) between the electrodes. A DC potential is used to drive the ions towards the surface of one of the electrodes (named the target) causing atoms to be knocked off the target and condense on the substrate surface. The process is typically performed on one side of the substrate at a time.

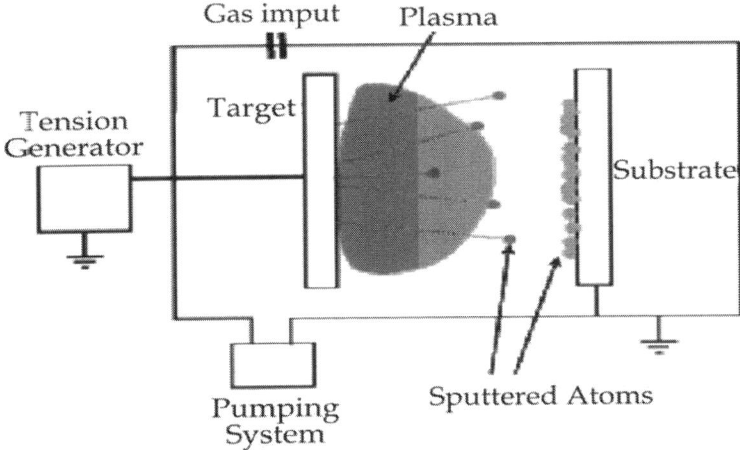

Fig. 9.4 : Radiofrequency sputtering

Magnetron Sputtering

When power is supplied to a magnetron a negative voltage of typically -300V or more is applied to the target. This negative voltage attracts positive ions to the target surface at speed. If the energy transferred to a lattice site is greater than the binding energy, primary recoil atoms can be created which can collide with other atoms and distribute their energy via collision cascades.A surface atom becomes sputtered if the energy transferred to it normal to the surface is larger than about 3 times the surface binding energy. Sputtering of the target atom is normally done by ion bombardment.

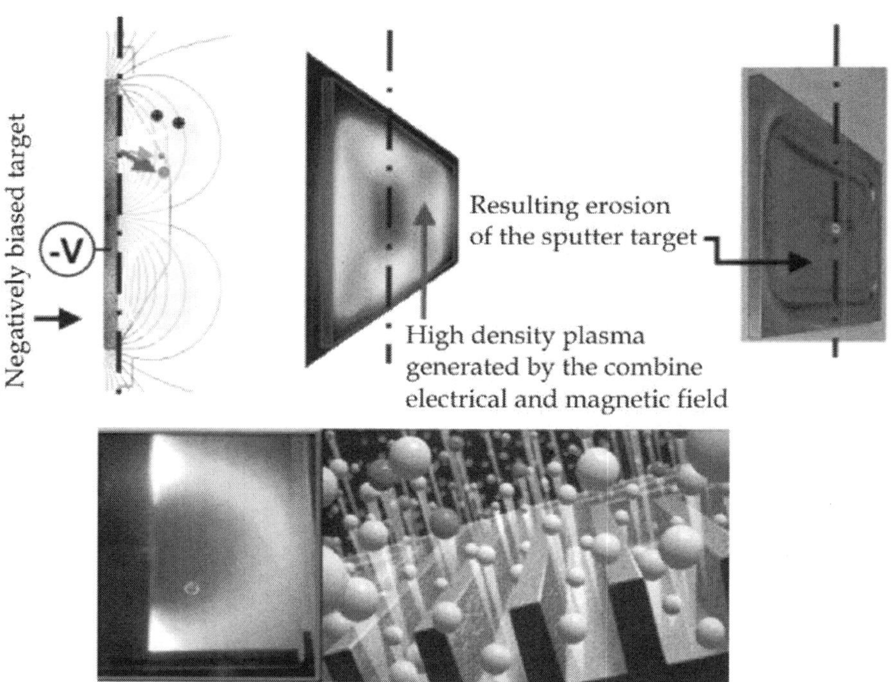

Fig. 9.5 : Magnetron sputtering

Pulsed Laser Deposition Technique

The process involves three stages. In the first stage, the laser beam is focused onto the surface of the target. The target elements are heated up to evaporation temperature at high flux densities and short pulse duration. Materials dissociated from the target surface and ablated out have the same stoichiometry as the target material.During the second stage the emitted materials move towards the substrate. The third stage is significant in determining the quality of thin film.

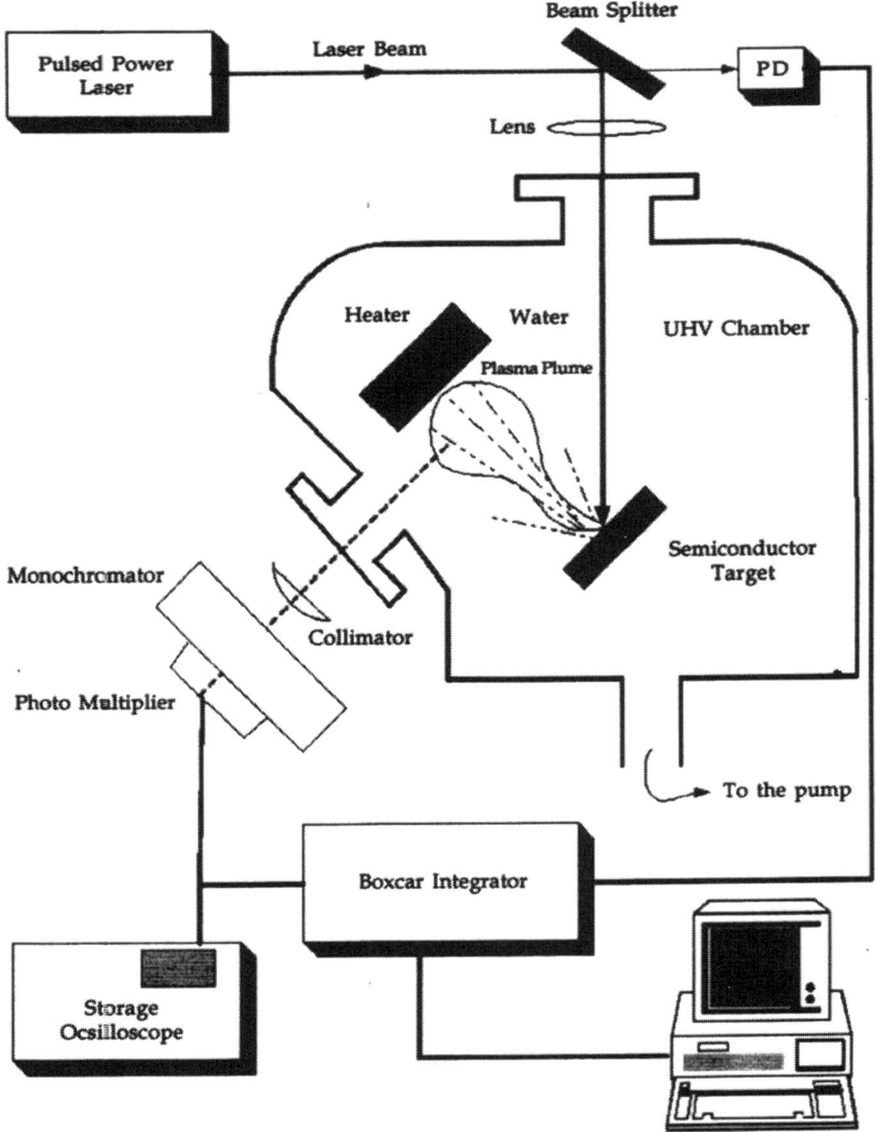

Fig. 9.6 : Pulsed Laser Deposition

Chemical Vapour Deposition (CVD)

In chemical vapor deposition (CVD), the substrate is placed inside a reaction vessel and the pressure and gas flow are controlled.The process involves a chemical reaction between source gases; the product of which condenses during the formation of a solid material within the reaction vessel.

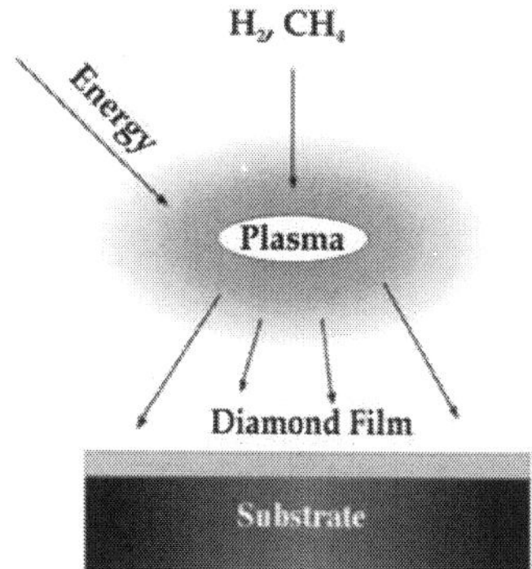

Fig. 9.7 : Chemical Vapour Deposition

Plasma-Enhanced Chemical Vapour Deposition

This method is commonly employed to deposit thin films from vapor state to a solid state on some substrate. The plasma is generally created by RF (AC) frequency or DC discharge between two electrodes, the space between which is filled with the reacting gases

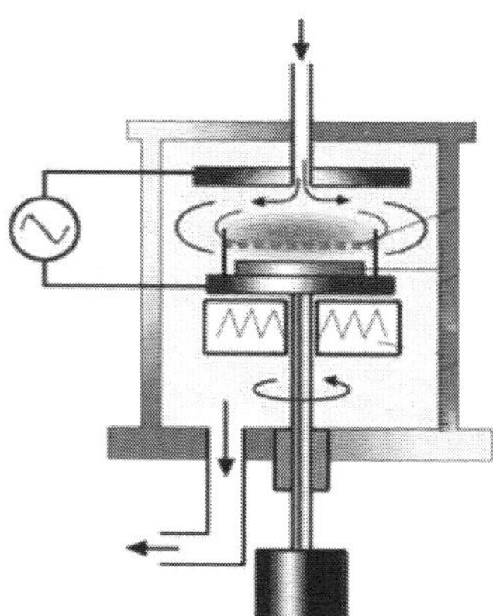

Fig. 9.8 : Plasma-Enhanced Chemical Vapour Deposition

Metal-Organic Chemical Vapor Deposition (MOCVD)

Chemical Vapour Deposition creates thin films of material on a substrate via the use of chemical reactions.Reactive gases are fed into a vacuum chamber and these gases react on a substrate and form a thin film or a powder. In this method the metal-organic precursor serves as a catalyst to carry out the chemical reaction.

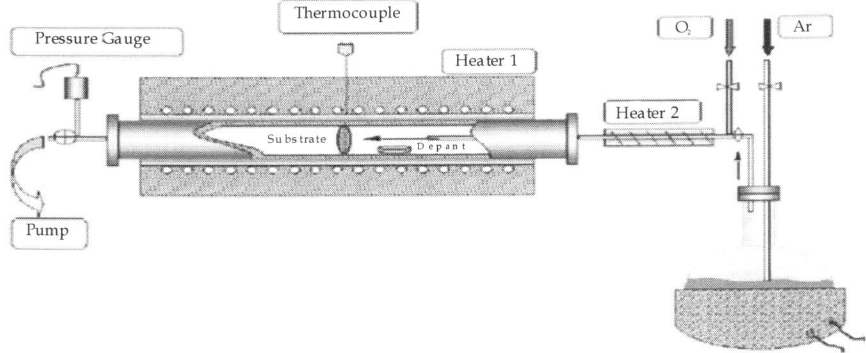

Fig. 9.9 : Metal-Organic Chemical Vapor Deposition

Molecular Beam Epitaxy

Molecular beam Epitaxy takes place in high vacuum.The most important aspect of MBE is the slow deposition rate (typically less than 1000 nm per minute), which allows the films to grow epitaxially.

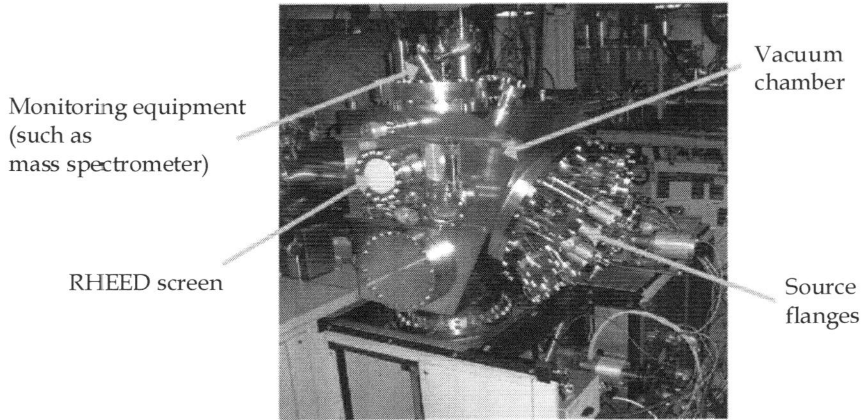

Fig. 9.10 : Molecular Beam Epitaxy

Atomic Layer Deposition (ALD)

It is a self-limiting, sequential, surface chemistry technique used to deposit thin films onto substrates with varying composition.It splits the chemical reaction into two half in order to prevent the precursors from interacting within the reaction vessel.

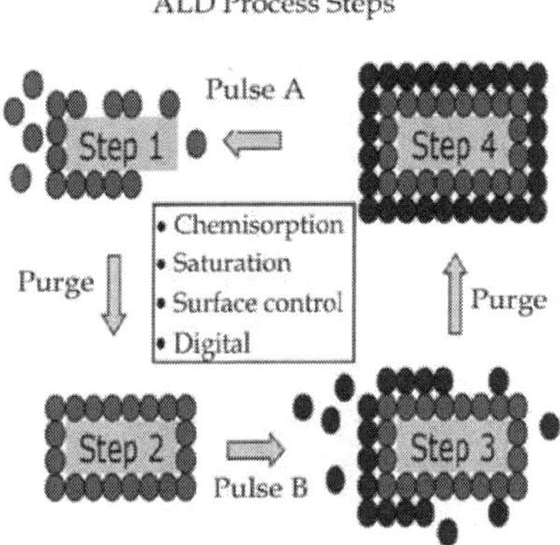

Fig. 9.11 : Atomic Longer Deposition

Synthesis of Nano Silicon by Laser Technique

Laser induced heating of silane to temperatures where it dissociates in the reactor shown. Silane absorbs laser energy at a wave length of 10.6 micro meters and the rate of production- 20-200 mg/h. It gives more yields.

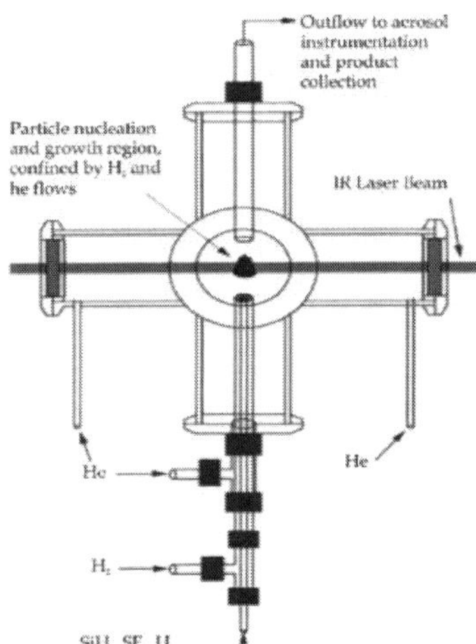

Fig. 9.12 : Laser Technique

Gas Condensation

Gas condensation was the first technique used to synthesize nanocrystalline metals and alloys. In this technique, a metallic or inorganic material is vaporized using thermal evaporation sources such as a Joule heated refractory crucibles, electron beam evaporation devices, in an atmosphere of 1-50 m bar. In gas evaporation, a high residual gas pressure causes the formation of ultra fine particles (100 nm) by gas phase collision. The ultrafine particles are formed by collision of evaporated atoms with residual gas molecules. Gas pressures greater than 3 mPa (10 torr) are required. Vaporization sources may be resistive heating, high energy electron beams, low energy electron beam and inducting heating. Clusters form in the vicinity of the source by homogenous nucleation in the gas phase grew by incorporation by atoms in the gas phase. It comprises of a ultra high vacuum (UHV) system fitted evaporation source, a cluster collection device of liquid nitrogen filled cold finger scrapper assembly and compaction device. During heating, atoms condense in the supersaturation zone close to Joule heating device. The nano particles are removed by scrapper in the form of a metallic plate. Evaporation is to be done from W, Ta or Mo refractory metal crucibles. If the metals react with crucibles, electron beam evaporation technique is to be used. The method is extremely slow. The method suffers from limitations such as a source-precursor incompatibility, temperature ranges and dissimilar evaporation rates in an alloy. Alternative sources have been developed over the years. For instance, Fe is evaporated into an inert gas atmosphere (He).Through collision with the atoms the evaporated Fe atoms loose kinetic energy and condense in the form of small crystallite crystals, which accumulate as a loose powder.

Sputtering or laser evaporation may be used instead of thermal evaporation. Sputtering is a non-thermal process in which surface atoms are physically ejected from the surface by momentum transfer from an energetic bombarding species of atomic/molecular size. Typical sputtering uses a glow discharge or ion beam. Interaction events which occur at and near the target surface during the sputtering process in magnetron sputtering has advantage over diode and triode sputtering. In magnetron sputtering, most of the plasma is confined to the near target region. Other alternate energy sources which have been successfully used to produce clusters or ultra fine particles are sputtering electron beam heating and plasma methods. Sputtering has been used in low pressure environment to produce a variety of clusters including Ag, Fe and Si.

Vacuum Deposition and Vaporization

Before proceeding to the other methods, it is important to understand the terms vacuum deposition and vaporization or vacuum evaporation. In vacuum deposition process, elements, alloys or compounds are vaporized and deposited in a vacuum .The vaporization source is the one that vaporizes materials by thermal processes. The process is carried out at pressure of less than 0.1 Pa (1 m Torr) and in vacuum levels of 10 to 0.1 MPa. The substrate temperature ranges from ambient to 500°C. The saturation or equilibrium vapor pressure of a material is defined as the vapor pressure of the material in equilibrium with the solid or liquid surface. For vacuum deposition, a reasonable deposition rate can be obtained if the vaporization rate is fairly high. A useful deposition rate is obtained at a vapor pressure of 1.3 Pa (0.01 Torr).

Vapor phase nucleation can occur in dense vapor cloud by multibody collisions. The atoms are passed through a gas to provide necessary collision and cooling for nucleation. These particles are in the range of 1 to 100 nm and are called ultra fine particles or clusters. The advantages associated with vacuum deposition process are high deposition rates and economy.

However, the deposition of many compounds is difficult. Nanoparticles produced from a supersaturated vapor are usually longer than the cluster.

Chemical Vapor Deposition (CVD) and Chemical Vapor Condensation (CVC)

CVD is a well known process in which a solid is deposited on a heated surface via a chemical reaction from the vapor or gas phase. CVC reaction requires activation energy to proceed.This energy can be provided by several methods. In thermal CVD the reaction is activated by a high temperature above 900C .A typical apparatus comprises of gas supply system, deposition chamber and an exhaust system. In plasma CVD, the reaction is activated by plasma at temperatures between 300 and 700°C.In laser CVD, pyrolysis occurs when laser thermal energy heats an absorbing substrate. In photo-laser CVD, the chemical reaction is induced by ultra violet radiation which has sufficient photon energy, to break the chemical bond in the reactant molecules.

In this process, the reaction is photon activated and deposition occurs at room temperature. Nano composite powders have been prepared by CVD. SiC/Si3N composite powder was prepared using SiH4, CH4, WF6 and H2 as a source of gas at 1400°C.It involves pyrolysis of vapors of metal organic precursors in a reduced pressure atmosphere. Particles of ZrO2, Y2O3 and nano whiskers have been produced by CVC method. A metalorganic precursor is introduced in the hot zone of the reactor using mass flow controller. For instance, hexamethyldisilazane (CH3)3 Si NHSi (CH3)3 was used to produce SiCxNyOz powder by CVC technique. The reactor allows synthesis of mixtures of nanoparticles of two phases or doped nanoparticles by supplying two precursors at the front end of reactor and coated nanoparticles, n-ZrO2, coated with n-Al2O3 by supplying a second precursor in a second stage of reactor.Another process called chemical vapor condensation (CVC) was developed in Germany in 1994.The process yields quantities in excess of 20 g/hr. The yield can be further improved by enlarging the diameter of hot wall reactor and mass of fluid through the reactor.

Mechanical Attrition

Unlike many of the methods mentioned above, mechanical attrition produces its nanostructures not by cluster assembly but by the structural decomposition of coarser grained structures as a result of plastic deformation. Elemental powders of Al and 2-SiC were prepared in a

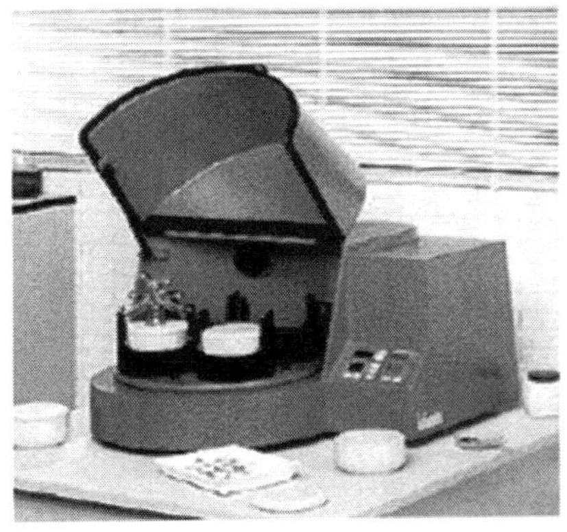

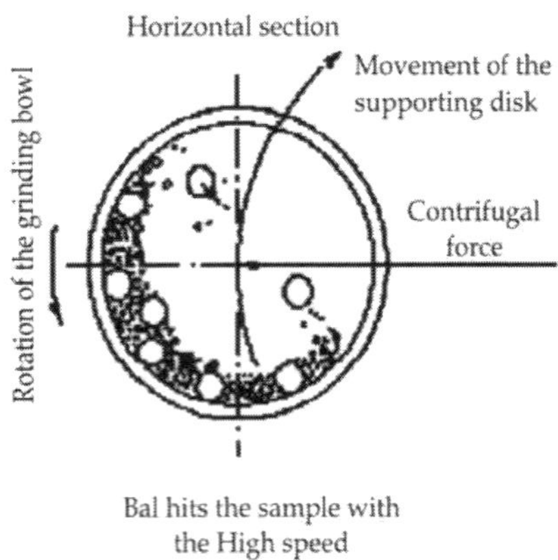

Fig. 9.13 : Mechanical Attrition

high energy ball mill. More recently, ceramic/ceramic nanocomposite WC-14% MgO material has been fabricated. The ball milling and rod milling techniques belong to the mechanical alloying process which has received much attention as a powerful tool for the fabrication of several advanced materials. Mechanical alloying is a unique process, which can be carried out at room temperature. The process can be performed on both high energy mills, centrifugal type mill and vibratory type mill, and low energy tumbling mill.

Examples of High Energy Mills

High energy mills include:

Attrition Ball Mill

Planetary Ball Mill

Vibrating Ball Mill

Low Energy Tumbling Mill

High Energy Ball Mill

Attrition Ball Mill

- The milling procedure takes place by a stirring action of a agitator which has a vertical rotator central shaft with horizontal arms (impellers).
- The rotation speed was later increased to 500 rpm.
- Also, the milling temperature was in greater control.

Planetary Ball Mill

- Centrifugal forces are caused by rotation of the supporting disc and autonomous turning of the vial.
- The milling media and charge powder alternatively roll on the inner wall of the vial and are thrown off across the bowl at high speed (360 rpm).

Vibrating Ball Mill

- It is used mainly for production of amorphous alloys.
- The changes of powder and milling tools are agitated in the perpendicular direction at very high speed (1200 rpm).

Low Energy Tumbling Mill

They have been used for successful preparation of mechanically alloyed powder.

They are simple to operate with low operation costs.

A laboratory scale rod mill was used to prepare homogenous amorphous $Al_{30}Ta_{70}$ powder by using S.S. cylinder rods.

Single-phase amorphous powder of AlxTm100-x with low iron concentration can be formed by this technique.

High Energy Ball Mill

High-energy ball milling is an already established technology, however, it has been considered dirty because of contamination problems with iron.However, the use of tungsten carbide component and inert atmosphere and /or high vacuum processes has reduced impurity levels to within acceptable limits.Common drawbacks include low surface, highly poly disperse size distribution, and partially amorphous state of the powder. These powders are highly reactive with oxygen, hydrogen and nitrogen. Mechanical alloying leads to the fabrication of alloys, which cannot be produced by conventional techniques. It would not be possible to produce an alloy of Al-Ta, because of the difference in melting points of Al (933 K) and Ta (3293 K) by any conventional process. However, it can be fabricated by mechanical alloying using ball milling process.

Other Processes

- Several other processes such as hydrodynamic cavitation micro emulsion and sono chemical processing techniques have also been used.
- In cavitation process nano particles are generated through creation and release of gas bubbles inside the sol-gel solution.
- By pressurizing in super critical drying chamber and exposing to cavitational disturbances and high temperature heating, the sol-gel is mixed.
- The erupted hydrodynamic bubbles cause the nucleation, growth and quenching of nano particles.
- Particle size can be controlled by adjusting pressure and solution retention times.

❑❑❑

Chapter-10

Chemical Methods of Nanoparticles Synthesis

Chemical synthesis is spontaneous organization of molecules into stable, structurally well-defined aggregates at the nanometer length scale. Chemical synthesis is largely an art form (i.e., theory often does not quantitatively predict how to make new materials from molecules).Yet, chemical synthesis can be quite powerful (e.g., proteins and complex organic molecules can be made from simple reagents) Semiconductor nanocrystals have been synthesized via "natural self-assembly" in two steps:

1. Capped cluster molecule forms spontaneously from molecular reagents in solution
2. Cluster molecules crystallize out of solution

Chemical methods

- Sol-gel methods
- Wet chemical co-precipitation
- Microencapsulation
- Hydrothermal methods

- Microwave synthesis
- Sonochemistry

Liquid Phase Synthesis Methods

The liquid phase fabrication entails a wet chemistry route.

Methods:

- Solvothermal Methods (e.g. hydrothermal)
- Sol-Gel Methods
- Synthesis in Structure Media (e.g., microemulsion)

Mechanism and Effectiveness

(i) Precursor solution (typically involves a catalyst)

(ii) Nucleation, and

(iii) Growth stage

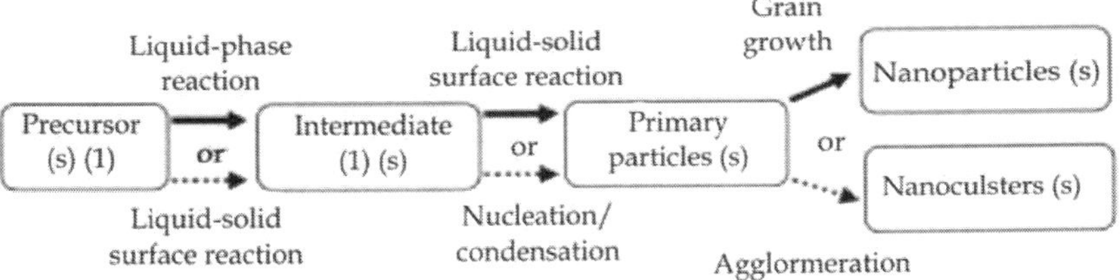

Fig 10.1 : Liquid Phase Fabrication

Effectiveness demands:

- Simple process
- Low cost Sol-Gel and Solvothermal Synthsis
- Continuous operation
- High yield

Sol-Gel Techniques

- The sol-gel processing techniques have also been extensively used.
- Colloidal particles are much larger than normal molecules or nanoparticles.

- However, upon mixing with liquid colloids appear bulky whereas the nanosized molecules always look clear.
- It involves the evolution of networks through the formation of colloidal suspension (sol) and gelatin to form a network in continuous liquid phase (gel).
- The precursor for synthesizing these colloids consists of ions of metal alkoxides and aloxysilanes.
- The most widely used are tetramethoxysilane (TMOS), and tetraethoxysilanes (TEOS) which form silica gels.
- Alkoxides are immiscible in water.
- They are organo metallic precursors for silica, aluminum, titanium, zirconium and many others. Mutual solvent alcohol is used.
- The sol gel process involves initially a homogeneous solution of one or more selected alkoxides. These are organic precursors for silica, alumina, titania, zirconia, among others. A catalyst is used to start reaction and control pH.

Sol-gel formation occurs in four stages.

- Hydrolysis
- Condensation
- Growth of particles
- Agglomeration of particles

Hydrolysis

- During hydrolysis, addition of water results in the replacement of [OR] group with [OH-] group.
- Hydrolysis occurs by attack of oxygen on silicon atoms in silica gel.
- Hydrolysis can be accelerated by adding a catalyst such as HCl and NH3.
- Hydrolysis continues until all alkoxy groups are replaced by hydroxyl groups.
- Subsequent condensation involving silanol group (Si-OH) produced siloxane bonds (Si-O-Si) and alcohol and water.

- Hydrolysis occurs by attack of oxygen contained in the water on the silicon atom.

Condensation

- Polymerization to form siloxane bond occurs by either a water producing or alcohol producing condensation reaction.
- The end result of condensation products is the formation of monomer, dimer, cyclic tetramer, and high order rings.
- The rate of hydrolysis is affected by pH, reagent concentration and H2O/Si molar ratio (in case of silica gels).
- Also ageing and drying are important.
- By control of these factors, it is possible to vary the structure and properties of sol-gel derived inorganic networks.

Growth and Agglomeration

- As the number of siloxane bonds increase, the molecules aggregate in the solution, where they form a network, a gel is formed upon drying.
- The water and alcohol are driven off and the network shrinks.
- At values of pH of greater then 7, and H2O/Si value ranging from 7 to 5.
- Spherical nano-particles are formed. Polymerization to form siloxane bonds by either an alcohol producing or water producing condensate occurs.

  $2\ HOSi\ (OR)_3 \rightarrow (OR)_3\ Si\ O\ Si\ (OR)_3 + H_2O$

  Or

  $2\ HOSi\ (OR)_3 \rightarrow (OR)2OH\ Si\ O\ Si\ (OR)_3 + H_2O$
- Above pH of 7, Silica is more soluble and silica particles grow in size.
- Growth stops when the difference in solubility between the smallest and largest particles becomes indistinguishable.
- Larger particles are formed at higher temperatures.
- Zirconium and Yttrium gels can be similarly produced.

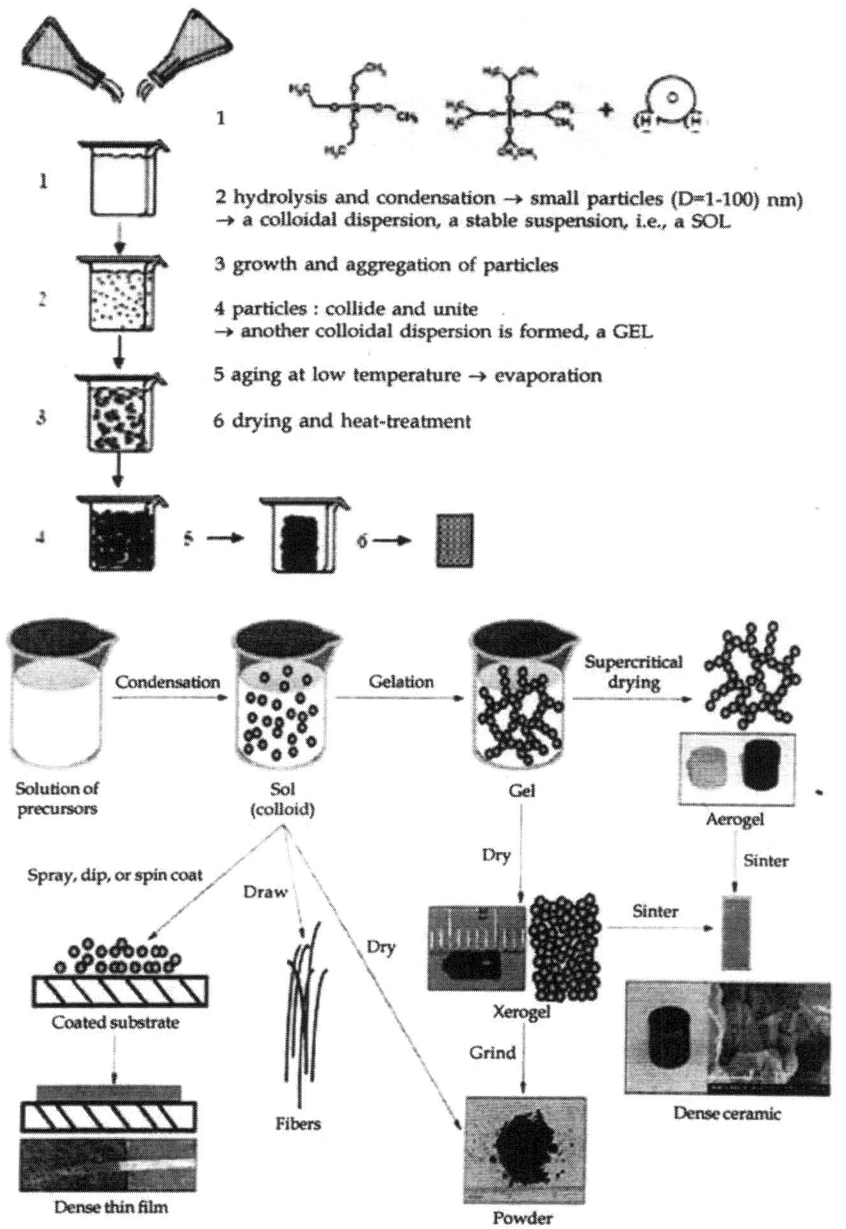

Fig. 10.2 : Sol-gel method

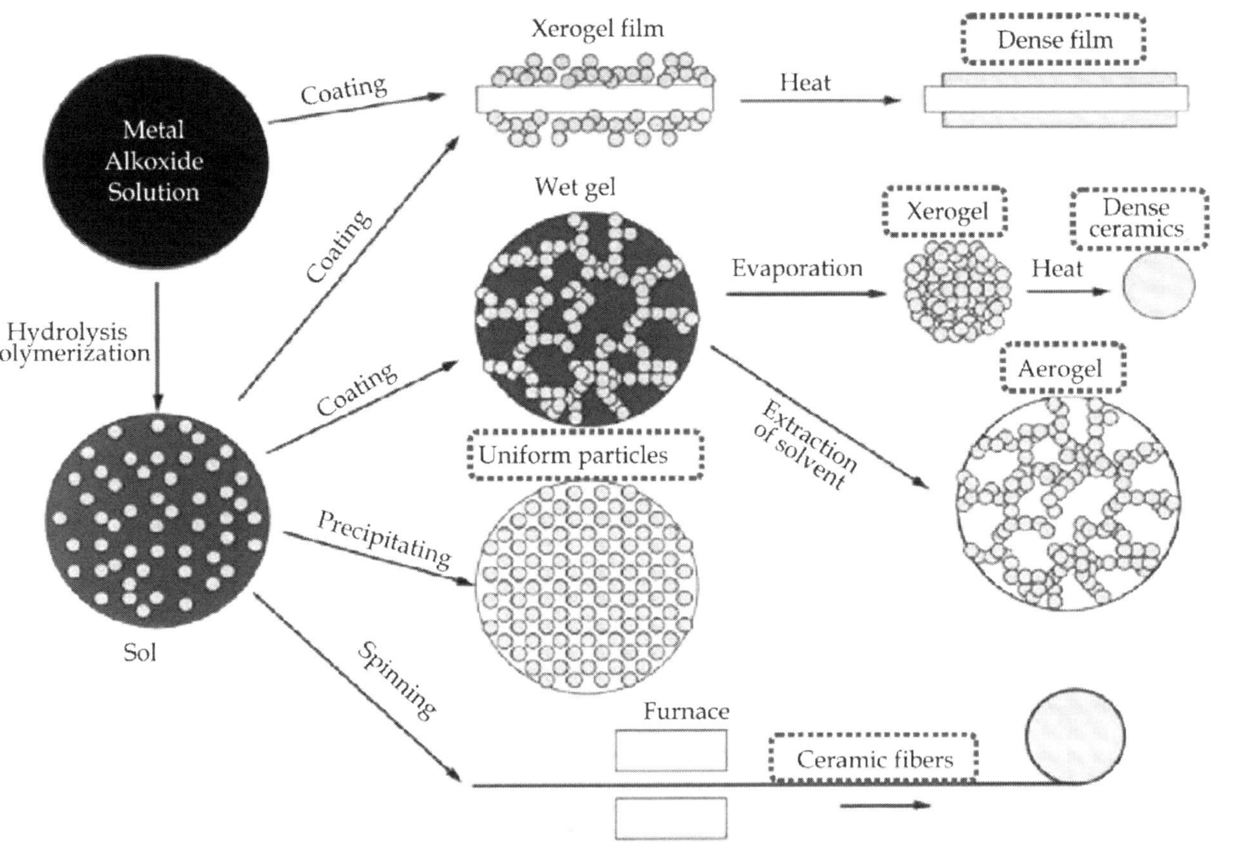

Fig. 10.3 : Sol-gel Techniques and their products

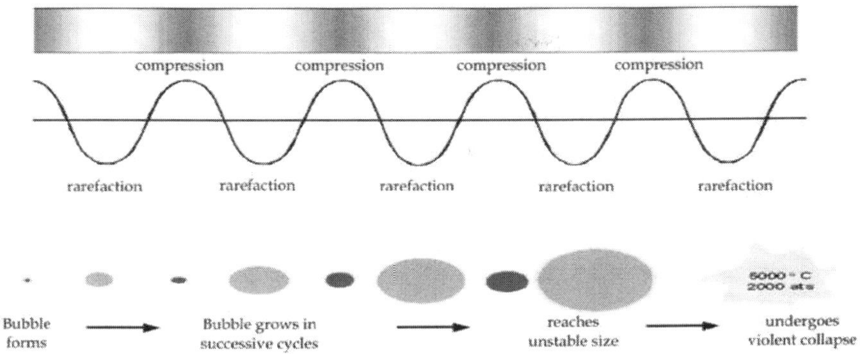

Fig. 10.4 : Acoustic cavitation

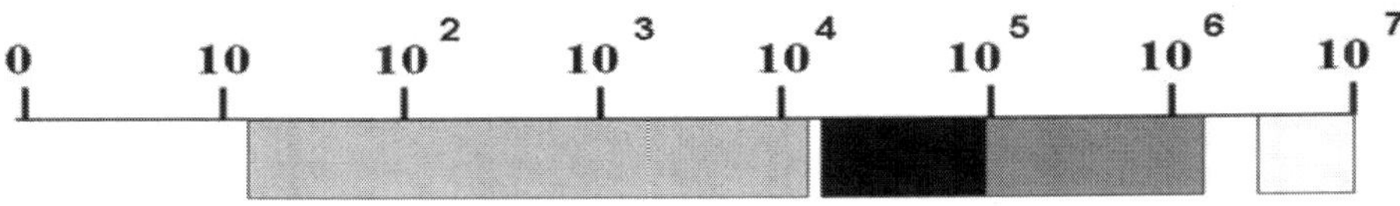

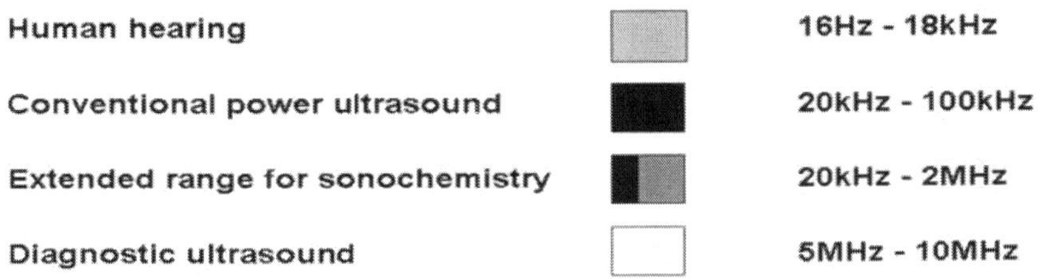

Fig. 10.5 : The Frequency Ranges of Sound

Sonochemistry

- Sonochemistry is the emerging study of chemical reactions powered by high-frequency sound waves.
- Ultrasonic waves in liquid cause the formation of tiny bubbles that collapse so quickly, and due to such enormous temperatures and pressures, those novel chemical reactions are generated.

- Chemical effects of ultrasound enhance reaction rates because of the formation of highly reactive radical species formed during cavitation in forms of bubbles.
- After being created, these bubbles are filled with vapor and gas, and can produce radicals during such an implosion.
- It is these implosions which are the spectacular part of sonochemistry.
- Each one of these imploding bubbles can therefore be seen as a micro reactor, with temperatures reaching an estimated 5000°C, and pressures of several hundreds of atmospheres.
- When ultrasound is supplied in the system, bubbles react with the compression/expansion cycle of ultrasound causing it to expand and collapse in compression , hence energy is released in term of light and heat at the point of implosion.
- The high temperatures and pressures created during cavitation can even cause the bubbles to produce a flash.
- Typically the bubbles are d riven belo w their natural frequency at high pressure amplitudes; the bubbles undergoslow expansions and rapid, catastrophic collapses.
- The bubble compression is so violent that the gas in the bubble has been estimated (through computations and experiments) to reach ~ 5 ,000- 8 000 K elvin and >10,000 atmospheres on a nanosecond time scale.

Sonochemical Approaches for Nano scale Particle Synthesis

Step 1:

- Bubble expands when surrounding medium experiences –ve pressure
- Bubble collapses when surrounding medium experiences +ve pressure
- Bubble collapse leads to extreme temperatures (5,000 – 50, 000 K), and pressure (100 atm) within the bubble

2: Step

- Solvent or solute molecules present within the bubbles are decomposed under these extreme conditions and generate highly reactive radicals

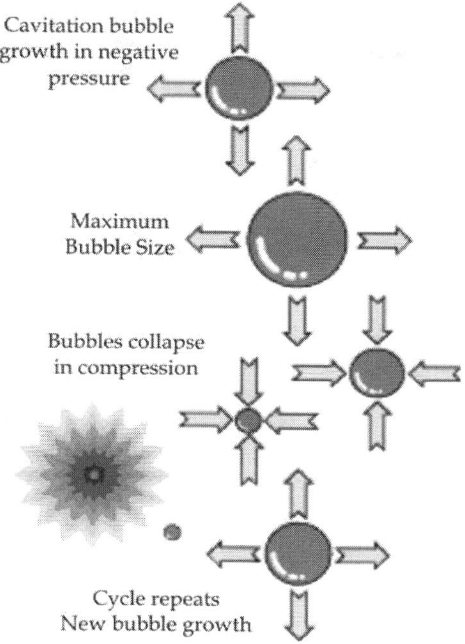

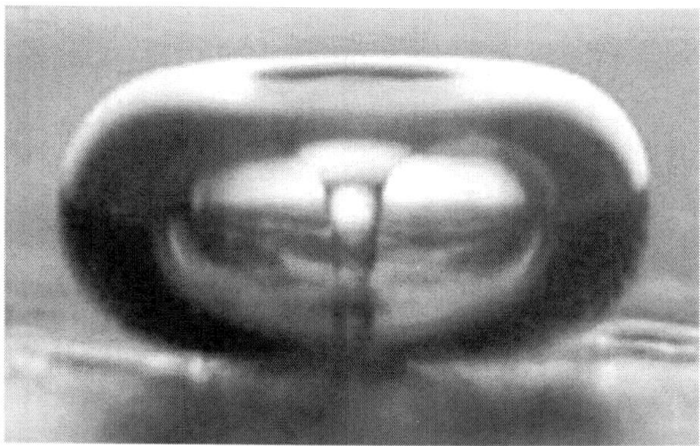

Fig. 10.6 : Creation, growth collapse of a bubble that is formed in the liquid

Solvothermal Synthesis

Precursors are dissolved in hot solvents (e.g., n-butyl alcohol)

- Solvent other than water can provide milder and friendlier reaction conditions. If the solvent is water then the process is referred to as hydrothermal method.

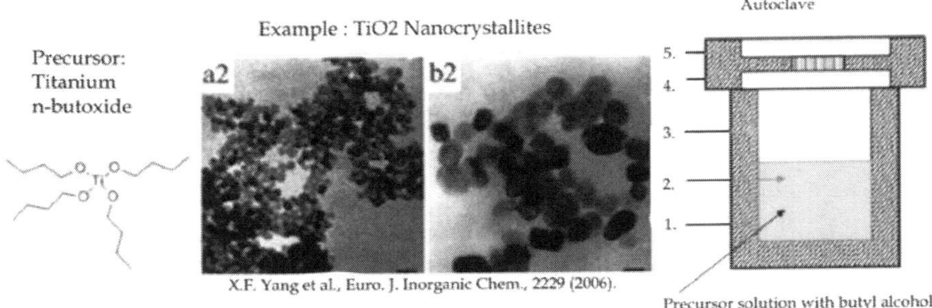

Solvothermal Synthesis

Solvothermal synthesis utilizes a solvent under pressures and temperatures above its critical point to increase the solubility of solid and to speed up reaction **between** solids

- Most materials can be made soluble in **proper solvent** by heating and pressuring the system close to its critical point
- Easily control a solubility of solute
  → Lower supersaturation state

1 Dimensional Nanostructure !

$$CdC_2O_4 + E \xrightarrow{\text{solvent}} CdE + 2CO_2 \uparrow \qquad E\text{ ; Chalcogenide (S, Se, Te)}$$

- Ethylenediamine
- Pyridine

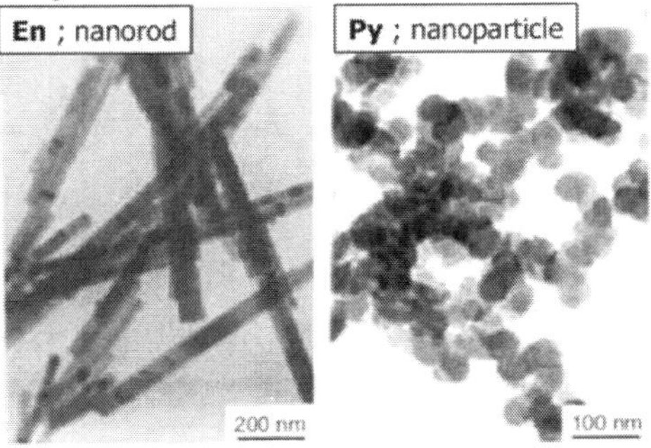

Fig. 10.8 : Semiconductor nanoparticles [CdE (E = S, Se, Te)]

Synthesis in Structured Medium

Influence Growth Kinetics by Imposing Constraints in Form of Matrices:

- Zeolites
- Layered Solids
- Molecular Sieves
- Micelles/Microemulsions
- Gels
- Polymers
- Glasses

Ex.: Mixing of two Micro emulsion carrying metal salt and reducing agent

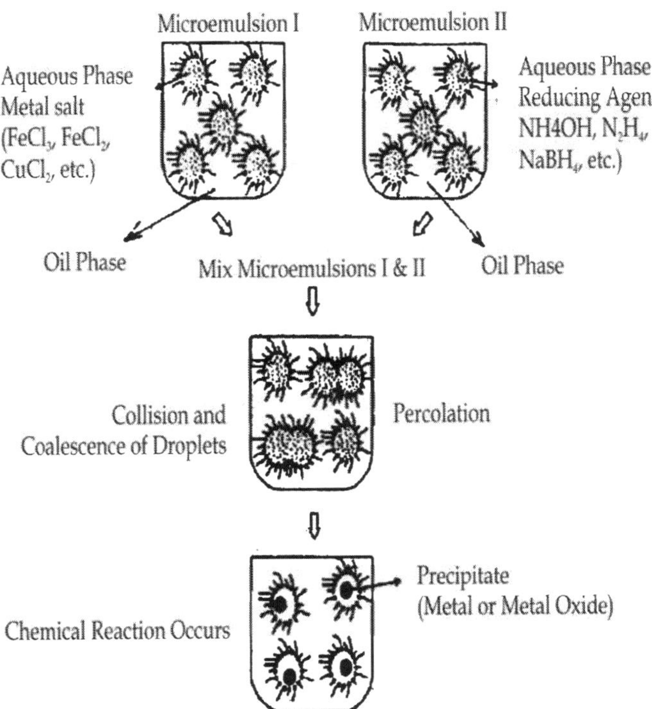

Fig. 10.9 : Inter micellar interchange process via coalescence (rate limiting) much slower than diffusion : 10 μs and 1 ms

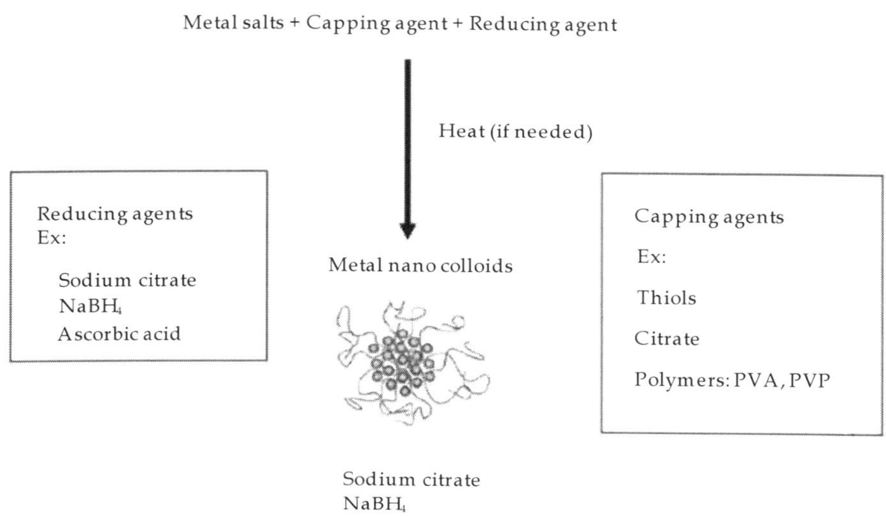

Fig. 10.10 : Synthesis of metal nanoparticles

1. Formation of 4 nm "seed"

2. Seed-mediated growth in the presence of (CTAB) rod-like Au spheroids & nanorods

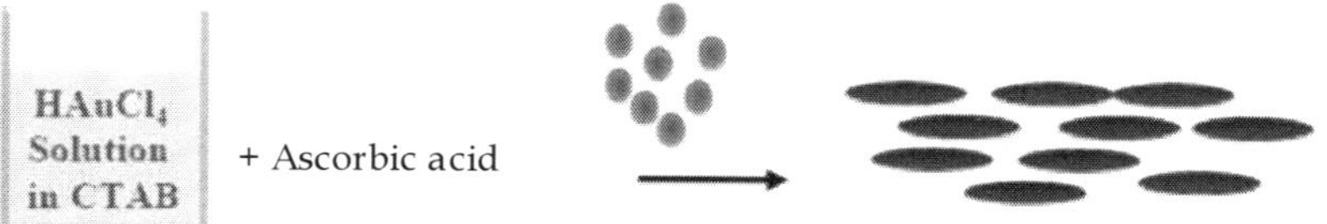

3. Seed-mediated growth Au nanorods

CTAB = cetyltrimethylammonium bromide (a surfactant)

Fig. 10.11 : Synthesis of Au Nanorods

Synthesis of Ag nanowires

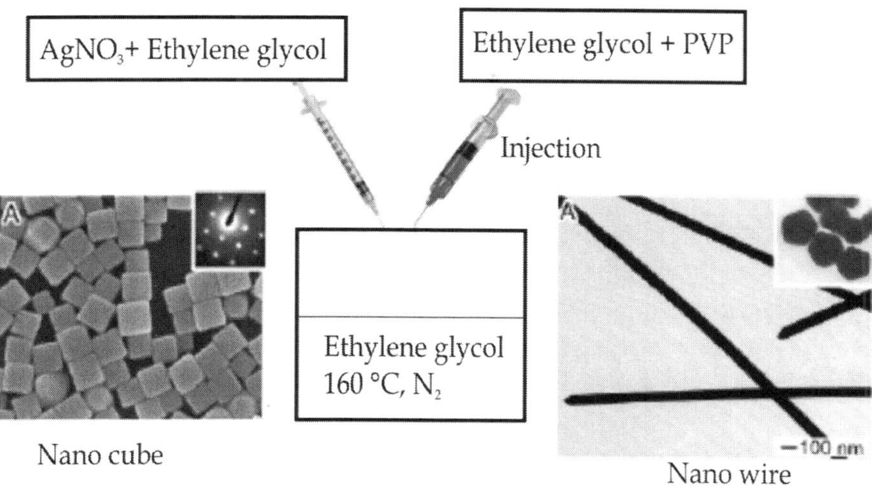

Fig. 10.12 : Polyol reduction method

Microwave-assisted synthesis

Role of microwave radiation:

Direct and very fast heating of the sample by microwave radiation induces uniform reactions, can improve the yield, and possibly enables different kinetic pathways of synthesis

Example: Synthesis of acicular magnetite

- Wet chemical synthesis: Acicular goethite (a-FeO(OH)) as starting material is transformed to hematite (acicular, a-Fe2O3), which is heated carefully under hydrogen to yield the product
- Microwave synthesis: Decomposition of Fe2+/Fe3+ aqueous solutions with concurrent dissociation of urea in a microwave field

Langmuir–Blodgett Film

- Langmuir–Blodgett film contains one or more monolayers of an organic material, deposited from the surface of a liquid onto a solid by immersing the solid substrate into (or from) the liquid.

- A monolayer is adsorbed homogeneously with each immersion or emersion step, thus films with very accurate thickness can be formed.
- This thickness is accurate because the thickness of each monolayer is known and can therefore be added to find the total thickness of a Langmuir-Blodgett Film.
- The monolayers are assembled vertically and are usually composed of amphiphilic molecules (see Chemical polarity) with a hydrophilic head and a hydrophobic tail (example: fatty acids).
- Alternative technique of creating single monolayers on surfaces is that of self-assembled monolayers.
- Langmuir–Blodgett films are named after Irving Langmuir and Katharine B. Blodgett, who invented this technique while working in Research and Development for General Electric Co.
- LB films are formed when amphiphilic molecules like surfactants interact with air at an air-water interface. Surfactants (or Surface acting agents) are molecules with hydrophobic 'tails' and hydrophilic 'heads'.
- When surfactant concentration is less than critical micellar concentration (CMC), the surfactant molecules arrange themselves as shown in Fig. 10.13.

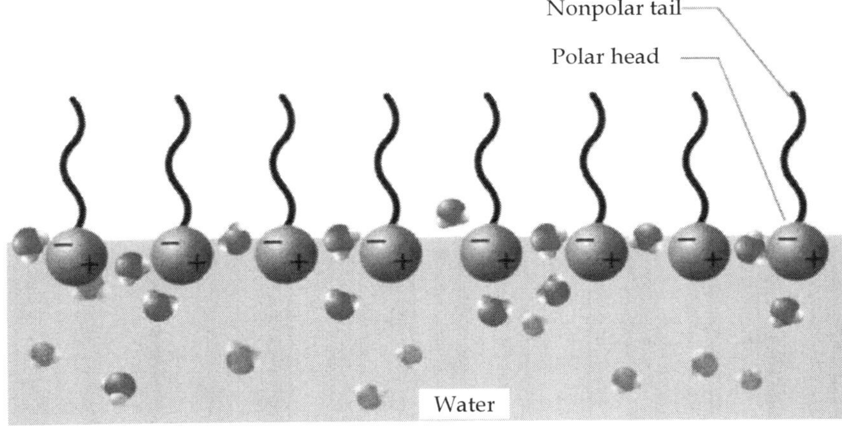

Fig. 10.13 : LB film

Surfactant molecules arranged on an air- water interface

- This tendency can be explained by surface-energy considerations.
- Since the tails are hydrophobic, their exposure to air is favoured over that to water.
- Similarly, since the heads are hydrophilic, the head-water interaction is more favourable than air-water interaction.
- The overall effect is reduction in the surface energy (or equivalently, surface tension of water).
- For very small concentrations, far less than critical micellar concentration (CMC), the surfactant molecules execute a random motion on the water-air interface.
- This motion can be thought to be similar to the motion of ideal gas molecules enclosed in a container.
- The corresponding thermodynamic variables for the surfactant system are, surface pressure ( ), surface area (A) and number of surfactant molecules (N).
- This system behaves similarly to a gas in a container. The density of surfactant molecules as well as the surface pressure increase upon reducing the surface area A ('compression' of the 'gas').
- Further compression of the surfactant molecules on the surface shows behavior similar to phase transitions.
- The 'gas' gets compressed into 'liquid' and ultimately into a perfectly closed packed array of the surfactant molecules on the surface corresponding to a 'solid' state.

❑❑❑

Chapter-11

Biological Methods of Nanoparticles Synthesis

Nanoparticles synthesis is done by two major routes, which include classical synthesis and green synthesis. The green synthesis techniques generally utilize relatively non-toxic chemicals non-toxic solvents, biological extracts and systems. Biological methods are considered safe and ecologically sound for the nanomaterial fabrication as an alternative to conventional physical and chemical methods

Nanoparticles

Gold, silver, and copper have been used mostly for the synthesis of stable dispersions of nanoparticles, which are useful in areas of photography, catalysis, biological labeling, photonics, optoelectronics and surface-enhanced Raman scattering (SERS) detection

The need for biosynthesis of nanoparticles rose as the physical and chemical processes were costly. So in the search of cheaper pathways for nanoparticle synthesis, scientists used microorganisms and then plant extracts for synthesis. Nature has devised various processes for the synthesis of nano- and micro- length scaled inorganic materials which have contributed to the development of relatively new and largely unexplored area of research based on the biosynthesis of nanomaterials.

Biosynthesis of nanoparticles is a kind of bottom up approach where the main reaction occurring is reduction/oxidation. The microbial enzymes or the plant phytochemicals with anti oxidant or reducing properties are usually responsible for reduction of metal compounds into their respective nanoparticles.

The three main steps in the preparation of nanoparticles that should be evaluated from a green chemistry perspective are the choice of the solvent medium used for the synthesis, the choice of an environmentally benign reducing agent and the choice of a non toxic material for the stabilization of the nanoparticles. Most of the synthetic methods reported to date rely heavily on organic solvents. This is mainly due to the hydrophobicity of the capping agents used. Synthesis using bio-organisms is compatible with the green chemistry principles: the bio-organism is (i) eco-friendly as are (ii) the reducing agent employed and (iii) the capping agent in the reaction. Often chemical synthesis methods lead to the presence of some toxic chemical species adsorbed on the surface that may have adverse effects in medical applications (Parashar *et al.*, 2009). This is not an issue when it comes to biosynthesized nano particles as they are eco friendly and biocompatible for pharmaceutical applications.

Use of organisms to synthesize nanoparticles

Biological route

Biological routes to the synthesis of these particles have been proposed by exploiting microorganisms and by vascular plants. The functions of these materials depend on theircomposition and structure. Plants have been reported to be used for synthesis of metal nanoparticles of gold and silver and of a gold-silver-copper alloy. Of this colloidal silver is of particular interest because of its distinctive properties such as good conductivity, chemical stability, and catalytic and antibacterial activity.

Biomolecules

Researchers report that biomolecules like protein, phenols and flavonoids not only play a role in reducing the ions to the nanosize, but also play an important role in the capping of the nanoparticles. The reduction of Ag^{+} ions by combinations of biomolecules found in these extracts such as vitamins, enzymes/proteins, organic acids such as citrates, amino acids, and polysaccharides is environmentally benign, yet chemically complex.

Biomimetics refers to applying biological principles for materials formation. One of the primary processes in biomimetics involves bioreduction. Initially bacteria were used to synthesize nanoparticles and this was later succeeded with the use of fungi, actinomycetes and more recently plants

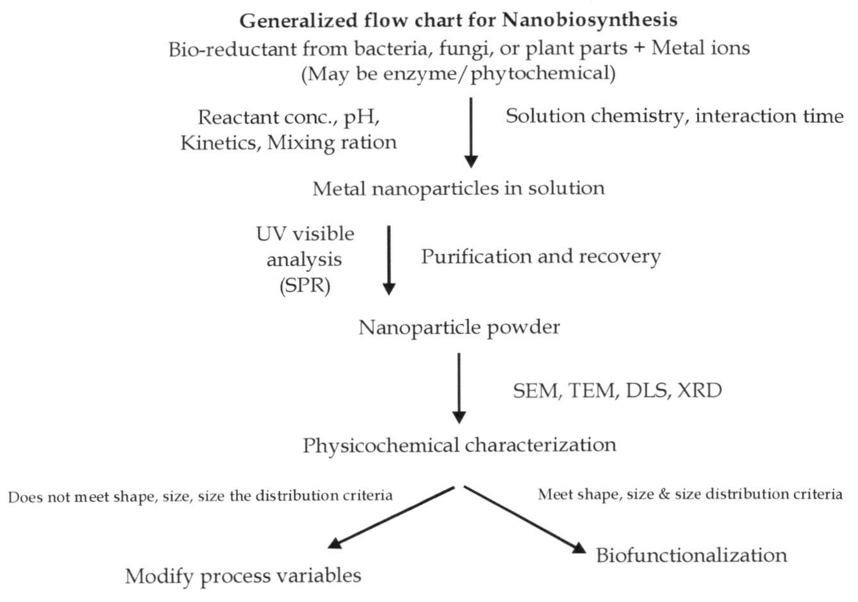

Fig. 11.1 : Flowchart denoting the biosynthesis of nanoparticles

Use of bacteria to synthesize nanoparticles

The use of microbial cells for the synthesis of nanosized materials has emerged as a novel approach for the synthesis of metal nanoparticles. Although the efforts directed towards the biosynthesis of nanomaterials are recent, the interactions between microorganisms and metals have been well documented and the ability of microorganisms to extract and/or accumulate metals is employed in commercial biotechnological processes such as bioleaching and bioremediation (Gericke & Pinches, 2006). Bacteria are known to produce inorganic materials either intracellularly or extra cellularly. Microorganisms are considered as a potential biofactory for the synthesis of nanoparticles like gold, silver and cadmium sulphide. Some well known examples of bacteria synthesizing inorganic materials include magnetotactic bacteria (synthesizing magnetic nanoparticles) and S layer bacteria which produce gypsum and calcium carbonate layers (Shankar *et al.*, 2004). Some microorganisms can survive and grow even at high metal ion concentration due to their resistance to the metal. The mechanisms involve: efflux systems, alteration of solubility and toxicity via reduction or oxidation, biosorption, bioaccumulation, extra cellular complexation or precipitation of metals and lack of specific metal transport systems (Husseiny *et al.*, 2007). For e.g. *Pseudomonas stutzeri* AG 259 isolated from silver mines has been shown to produce silver nanoparticles (Mohanpuria *et al.*, 2007). Many microorganisms are known to produce nanostructured mineral crystals and metallic nanoparticles with properties similar to chemically synthesized materials, while exercising strict control over size, shape and composition of the particles. Examples include the formation of magnetic nanoparticles by magnetotactic bacteria, the production of silver nanoparticles within the periplasmic space of *Pseudomonas stutzeri* and the formation of palladium nanoparticles using sulphate reducing bacteria in the presence of an exogenous electron donor (Gericke & Pinches, 2006).

Though it is widely believed that the enzymes of the organisms play a major role in the bioreduction process, some studies have indicated it otherwise. Studies indicate that some microorganisms could reduce silver ions where the processes of bioreduction were probably non enzymatic. For e.g. dried cells of *Bacillus megaterium* D01, *Lactobacillus sp.* A09 were shown to reduce silver ions by the interaction of the silver ions with the groups on the microbial cell wall (Fu *et al.*, 1999, 2000). Silver nanoparticles in the size range of 10- 15 nm were produced by treating dried cells of *Corynebacterium sp.* SH09 with diammine silver complex. The ionized carboxyl group of amino acid residues and the amide of peptide chains were the main groups trapping (Ag(NH3)2+) onto the cell wall and some reducing groups such as aldehyde and ketone were involved in subsequent bioreduction. But it was found that the reaction progressed slowly and could be accelerated in the presence of OH-(Fu *et al.*, 2006).In the case of bacteria, most metal ions are toxic and therefore the reduction of ions or the formation of water insoluble complexes is a defense mechanism developed by the bacteria to overcome such toxicity (Sastry *et al.*, 2003).

Use of actinomycetes to synthesize nanoparticles

Actinomycetes are microorganisms that share important characteristics of fungi and prokaryotes such as bacteria. Even though they are classified as prokaryotes, they were originally designated as ray fungi. Focus on actinomycetes has primarily centred on their exceptional ability to produce secondary metabolites such as antibiotics. It has been observed that a novel alkalothermophilic actinomycete, *Thermomonospora sp.*synthesized gold nanoparticles extracellularly when exposed to gold ions under alkaline conditions (Sastry *et al.*, 2003). In an effort to elucidate the mechanism or the processes favouring the formation of nanoparticles with desired features, Ahmad *et al.* (2003), studied the formation of monodisperse gold nanoparticles by *Thermomonospora sp.* and concluded that extreme biological conditions such as alkaline and slightly elevated temperature conditions were favourable for the formation of monodisperse particles. Based on this hypothesis, alkalotolerant actinomycete *Rhodococcus sp.* has been used for the intracellular synthesis of monodisperse gold nanoparticles by Ahmad *et al.* (2003). In this study it was observed that the concentration of nanoparticles were more on the cytoplasmic membrane. This could have been due to the reduction of metal ions by the enzymes present in the cell wall and on the cytoplasmic membrane but not in the cytosol. The metal ions were also found to be non toxic to the cells which continued to multiply even after the formation of the nanoparticles.

Use of fungi to synthesize nanoparticles

Fungi have been widely used for the biosynthesis of nanoparticles and the mechanistic aspects governing the nanoparticle formation have also been documented for a few of them. In addition to monodispersity, nanoparticles with well defined dimensions can be obtained using fungi. Compared to bacteria, fungi could be used as a source for the production of large amount of nanoparticles. This is due to the fact that fungi secrete more amounts of proteins which directly translate to higher productivity of nanoparticle formation (Mohanpuria *et al.*, 2007).

Yeast, belonging to the class ascomycetes of fungi has shown to have good potential for the synthesis of nanoparticles. Gold nanoparticles have been synthesized intracellularly using the fungi *V.luteoalbum*. Here, the rate of particle formation and therefore the size of the nanoparticles could to an extent be manipulated by controlling parameters such as pH, temperature, gold concentration and exposure time. A biological process with the ability to strictly control the shape of the particles would be a considerable advantage (Gericke & Pinches, 2006).

Extracellular secretion of the microorganisms offers the advantage of obtaining large quantities in a relatively pure state, free from other cellular proteins associated with the organism with relatively simpler downstream processing. Mycelia free spent medium of the fungus, *Cladosporium cladosporioides* was used to synthesise silver nanoparticles extracellularly. It was hypothesized that proteins, polysaccharides and organic acids released by the fungus were able to differentiate different crystal shapes and were able to direct their growth into extended spherical crystals (Balaji *et al.*, 2009). *Fusarium oxysporum* has been reported to synthesize silver nanoparticles extracellularly. Studies indicate that a nitrate reductase was responsible for the reduction of silver ions and the corresponding formation of silver nanoparticles. However *Fusarium moniliformae* did not produce nanoparticles either intracellularly or extracellularly even though they had intracellular and extracellular reductases in the same fashion as *Fusarium oxysporum*. This indicates that probably the reductases in *F.moniliformae* were necessary for the reduction of Fe (III) to Fe (II) and not for Ag (I) to Ag (0) (Duran *et al.*, 2005).

Instead of fungi culture, isolated proteins from them have also been used successfully in nanoparticles production. Nanocrystalline zirconia was produced at room temperature by cationic proteins while were similar to silicatein secreted by *F. oxysporum* (Mohanpuria *et al.*, 2007).

The use of specific enzymes secreted by fungi in the synthesis of nanoparticles appears promising. Understanding the nature of the biogenic nanoparticle would be equally important. This would lead to the possibility of genetically engineering microorganisms to over express specific reducing molecules and capping agents and thereby control the size and shape of the biogenic nanoparticles (Balaji *et al.*, 2009).

Microbiological methods generate nanoparticles at a much slower rate than that observed when plant extracts are used. This is one of the major drawbacks of biological synthesis of nanoparticles using microorganisms and must be corrected if it must compete with other methods.

Use of plants to synthesize nanoparticles

The advantage of using plants for the synthesis of nanoparticles is that they are easily available, safe to handle and possess a broad variability of metabolites that may aid in reduction.

A number of plants are being currently investigated for their role in the synthesis of nanoparticles. Gold nanoparticles with a size range of 2- 20 nm have been synthesized using the live alfa alfa plants (Torresday *et al.*, 2002). Nanoparticles of silver, nickel, cobalt, zinc and copper have also been synthesized inside the live plants of *Brassica juncea* (Indian mustard), *Medicago sativa* (Alfa alfa) and *Heliantus annus* (Sunflower). Certain plants are known to accumulate higher concentrations of metals compared to others and such plants are termed as hyperaccumulators. Of the plants investigated, *Brassica juncea* had better metal accumulating ability and later assimilating it as nanoparticles (Bali *et al.*, 2006). Recently much work has been done with regard to plant assisted reduction of metal nanoparticles and the respective role of phytochemicals. The main phytochemicals responsible have been identified as terpenoids, flavones, ketones, aldehydes, amides and carboxylic acids in the light of IR spectroscopic studies. The main water soluble phytochemicals are flavones, organic acids and quinones which are responsible for immediate reduction. The phytochemicals present in *Bryophyllum sp.* (Xerophytes), *Cyprus sp.* (Mesophytes) and *Hydrilla sp.* (Hydrophytes) were studied for their role in the synthesis of silver nanoparticles. The Xerophytes were found to contain emodin, an anthraquinone which could undergo redial tautomerization leading to the formation of silver nanoparticles. The Mesophyte studied contained three types of benzoquinones, namely, cyperoquinone, dietchequinone and remirin. It was suggested that gentle warming followed by subsequent incubation resulted in the activation of quinones leading to particle size reduction. Catechol and protocatechaldehyde were reported in the hydrophyte studied along with other phytochemicals. It was reported that catechol under alkaline conditions gets transformed into protocatechaldehyde and finally into protocatecheuic acid. Both these processes liberated hydrogen and it was suggested that it played a role in the synthesis of the nanoparticles. The size of the nanoparticles synthesized using xerophytes, mesophytes and hydrophytes were in the range of 2- 5nm (Jha *et al.*, 2009).

Recently gold nanoparticles have been synthesized using the extracts of *Magnolia kobus* and *Diopyros kaki* leaf extracts. The effect of temperature on nanoparticle formation was investigated and it was reported that polydisperse particles with a size range of 5- 300nm was obtained at lower temperature while a higher temperature supported the formation of smaller and spherical particles (Song *et al.*, 2009).

While fungi and bacteria require a comparatively longer incubation time for the reduction of metal ions, water soluble phytochemicals do it in a much lesser time. Therefore compared to bacteria and fungi, plants are better candidates for the synthesis of nanoparticles. Taking use of plant tissue culture techniques and downstream processing procedures, it is possible to synthesize metallic as well as oxide nanoparticles on an industrial scale once issues like the metabolic status of the plant etc. are properly addressed.

Work on the biomimetic synthesis of nanoparticles in India

There has been considerable significant research in India in the field of biomimetic synthesis of nanoparticles. More research has been found to be concentrated in the area of biomimetic synthesis using plants.

It has been observed that a novel alkalothermophilic actinomycete, *Thermomonospora sp.* synthesized gold nanoparticles extracellularly when exposed to gold ions under alkaline conditions (Sastry *et al.*, 2003). The use of algae for the biosynthesis of nanoparticles is a largely unexplored area. There is very little literature supporting its use in nanoparticle formation. Recently stable gold nanoparticles have been synthesized using the marine alga, *Sargassum wightii*. Nanoparticles with a size range between 8nm to 12nm were obtained using the seaweed. An important potential benefit of the method of synthesis was that the nanoparticles were quite stable in solution (Singaravelu *et al.*, 2007). Yeast, belonging to the class ascomycetes of fungi has shown to have good potential for the synthesis of nanoparticles. *Schizosaccharomyces pombe* cells were found to synthesize semiconductor CdS nanocrystals and the productivity was maximum during the mid log phase of growth. Addition of Cd in the initial exponential phase of yeast growth affected the metabolism of the organism (Kowshik *et al.*, 2002). Baker's yeast (*Saccharomyces cerevisiae*) has been reported to be a potential candidate for the transformation of Sb_2O_3 nanoparticles and the tolerance of the organism towards Sb_2O_3 has also been assessed. Particles with a size range of 2- 10 nm were obtained.

Aspergillus flavus has been found to accumulate silver nanoparticles on the surface of its cell wall when challenged with silver nitrate solution. Monodisperse silver nanoparticles with a size range of 8.92+/- 1.61nm were obtained and it was also found that a protein from the fungi acted as a capping agent on the nanoparticles (Vigneshwaran *et al.*, 2007). *Aspergillus fumigatus* has been studied as a potential candidate for the extracellular biosynthesis of silver nanoparticles. The advantage of using this organism was that the synthesis process was quite rapid with the nanoparticles being formed within minutes of the silver ion coming in contact with the cell filtrate. Particles with a size range of 5- 25nm could be obtained using this organism (Bhainsa & D Souza, 2006).

In addition to the synthesis of silver nanoparticles, *Fusarium oxysporum* has also been used to synthesize zirconia nanoparticles. It has been reported that cationic proteins with a molecular weight of 24-28 kDa (similar in nature to silicatein) were responsible for the synthesis of the nanoparticles (Bansal *et al.*, 2004).Recently, scientists in India have reported the green synthesis of silver nanoparticles using the leaves of the obnoxious weed, *Parthenium hysterophorus*. Particles in the size range of 30- 80nm were obtained after 10 min of reaction. The use of this noxious weed has an added advantage in that it can be used by nanotechnology processing industries (Parashar *et al.*,2009). *Mentha piperita* leaf extract has also been used recently for the synthesis of silver nanoparticles. Nanoparticles in the size range of 10-25 nm were obtained within 15 min of the reaction (Parashar *et al.*, 2009).

Azadirachta indica leaf extract has also been used for the synthesis of silver, gold and bimetallic (silver and gold) nanoparticles. Studies indicated that the reducing phytochemicals in the neem leaf consisted mainly of terpenoids. It was found that these reducing components also served as capping and stabilizing agents in addition to reduction as revealed from FT IR studies. The major advantage of using the neem leaves is that it is a commonly available medicinal plant and the antibacterial activity of the biosynthesized silver nanoparticle might have been enhanced as it was capped with the neem leaf extract.

The major chemical constituents in the extract were identified as nimbin and quercetin (Shankar *et al.*, 2004, Tripathy *et al.*, 2009). Figure 2 and 3 show the TEM micrograph of the biosynthesized silver nanoparticles (unpublished data, Prathna T.C. *et al.*, 2009).

Green synthesis of silver nanoparticles –*Desmodium*

Mechanism

The mechanism for the reduction of Ag ions to silver could be due to the presence of water-soluble antioxidative substances like ascorbate. This acid is present at high levels in all parts of plants. Ascorbic acid is a reducing agent and can reduce, and thereby neutralize, reactive oxygen species leading to the formation of ascorbate radical and an electron. This free electron reduces the Ag+ ion to Ag0.

It has been reported that ionic silver strongly interacts with thiol group of vital enzymes and inactivates them. Experimental evidence suggests that DNA loses its replication ability once the bacteria have been treated with silver ions. The antibacterial effect of nanoparticles can be attributed to their stability in the medium as a colloid, which modulates the phosphotyrosine profile of the bacterial proteins and arrests bacterial growth.

Fig. 11.2 : Green synthesis of NP from Desmodium

Synthesis from herb

Indian researchers at Patna University have biosynthesised silver nanoparticles from Desmodium triflorum. Desmodium triflorum is a wild much branched slender diffused herb with trifoliate leaves occurring as small under herb found in grasslands, fields, and agricultural lands forming a green turf on the ground. The dry plant was powdered, added with distilled water, heated and the extract was added to AgNO3 solutions. The bioreduction of Ag+ ions took place . The solution containing the signatory color of AgNPs (dark brown) was dryed in oven to get powders of silver nanoparticles. Thus stable and spherically shaped nanoparticles of average size ~10nm were synthesized using desmodium plant. The green synthesis of AgNPs fulfills all the three main steps, which must be evaluated based on green chemistry perspectives, including selection of solvent medium, selection of environmentally benign reducing agent and selection of nontoxic substances for the AgNPs stability. The study further showed that Ag nanoparticles presented good antibacterial performance against common pathogens. The nanoparticles when combined with the antibiotics show synergic effect in suppressing growth of antibiotics.

.Green Synthesis of Gold Nanoparticles

A simple, green, one-pot synthesis of gold nanoparticles (AuNPs) was achieved by Kurt E. Geckeler, Gwangju Institute of Science and Technology (GIST), and Thathan Premkumar, World-Class University (WCU), Gwangju, South Korea, through the reaction of an aqueous mixture of potassium tetrachloroaurate(III) and the macrocycle cucurbit[7]uril in the presence of sodium hydroxide at room temperature without introducing any kind of traditional reducing agents and/or external energy.

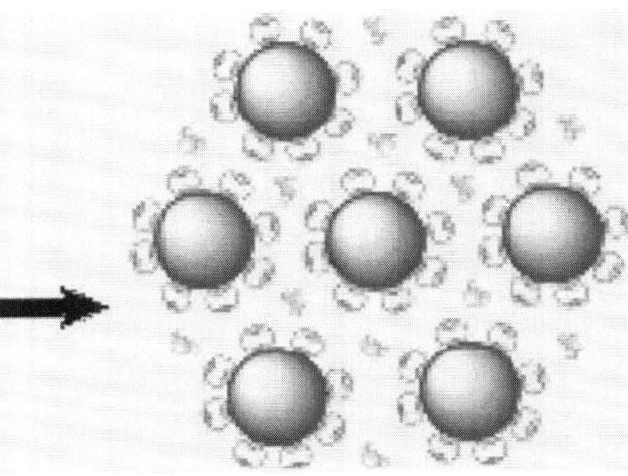

Fig. 11.3 : Gold nanoparticles

The as-prepared gold nanoparticles showed catalytic activity for the reduction reaction of 4-nitrophenol in the presence of $NaBH_4$, which has been established by visual inspection and UV/Vis spectroscopy.

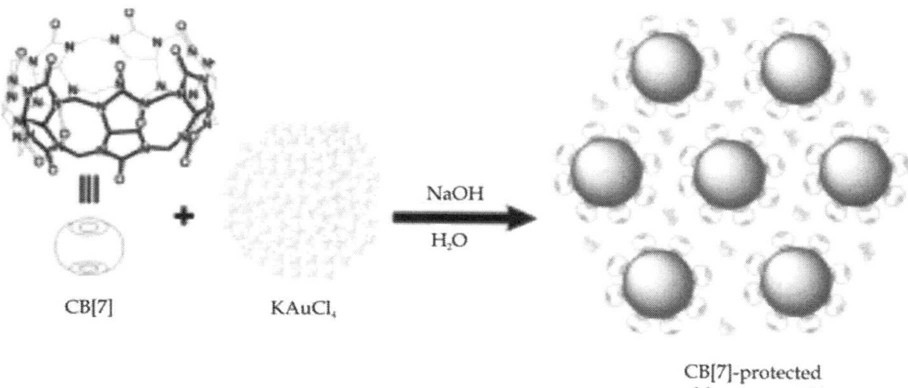

Surprisingly, the macrocycle can play a dual role in the synthesis of AuNPs: as a reducing agent as well as being a protecting agent.

This report is the first for the preparation of gold nanoparticles using cucurbit[7]uril in aqueous media through chemical reduction without employing conventional reducing agents and/or external energy.

Green synthesis of silver nanoparticles

Silver nanoparticles have extensive application in the development of new technologies in the areas of electronics, medicine and material sciences due to good conductivity and chemical stability. For example

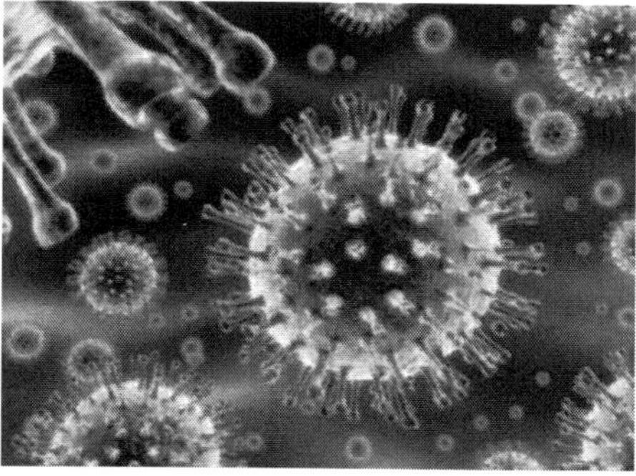

Fig. 11.4 : Silver nanoparticles

they find wide application as spectrally selective coatings for solar energy absorption, intercalation material for electrical batteries, optical receptors, catalysts in chemical reactions, bio labeling, photonics, optoelectronics, surface-enhanced Raman scattering (SERS) detection material and as antimicrobials.

Silver nano-particles

Many reports are available on the biogenesis of silver nanoparticles using several plant extracts, particularly neem leaf broth (Azadirachta indica), pelargonium graveolens (P.graveolens- geranium), geranium leaves (P. graveolens), medicago sativa (Alfalfa), aloe vera, emblica officinalis (amla, Indian Gooseberry) and few microorganisms. Similarly different plant constituents such as geraniol possess reducing property and reduce Ag+ to silver nanoparticles with a uniform size and shape in the range of 1 to 10 nm with an average size of 6 nm.

Factors for biosynthesis

Indian researchers have employed a single-step environmental friendly approach to synthesize silver nanoparticles. According to their report the biomolecules found in plants induce the reduction of Ag+ ions from silver nitrate to silver nanoparticles (AgNPs). The process of reduction is extra cellular and fast leading to the development of easy biosynthesis of silver nanoparticles. Plants during glycolysis produce a large amount of H+ ions along with NAD which acts as a strong redoxing agent which is responsible for the formation of AgNPs. Water-soluble antioxidative agents like ascorbic acids are further responsible for the reduction of AgNPs. These AgNPs produced show good antimicrobial activity against common pathogens.

The polyol components and the water-soluble heterocyclic components are largely accountable for the reduction of Ag ions and the stabilization of the nanoparticles, respectively. There are also reports on reductases and polysaccharides as factors involved in biosynthesis and stabilization of the nanoparticles, respectively. Also the sunflower leaf extract was found to be promising in the development of silver nanoparticles.Recently, efficient antibacterial activity was observed against multi drug resistant and highly pathogenic bacteria, including multi drug resistant Staphylococcus aureus, Salmonella typhi, Staphylococcus epidermidis and Escherichia coli by silver nanoparticles produced by the fungus F. acuminatum.

However, the synthesis of silver nanoparticles using such plant constituents or microorganisms has not yet been studied for a large number of natural compounds.

Microwave-assisted gold nanoparticle biosynthesis – A clean-green technology

In the new venture of green chemical technology Indian scientists discovered that the gold nanoparticles may be biosynthesized in next to no time. They have used microwave-assisted anti-malignant plant leaf of guava (Psidium guajava) for the biosynthesis of stable polyshaped gold nanoparticles.

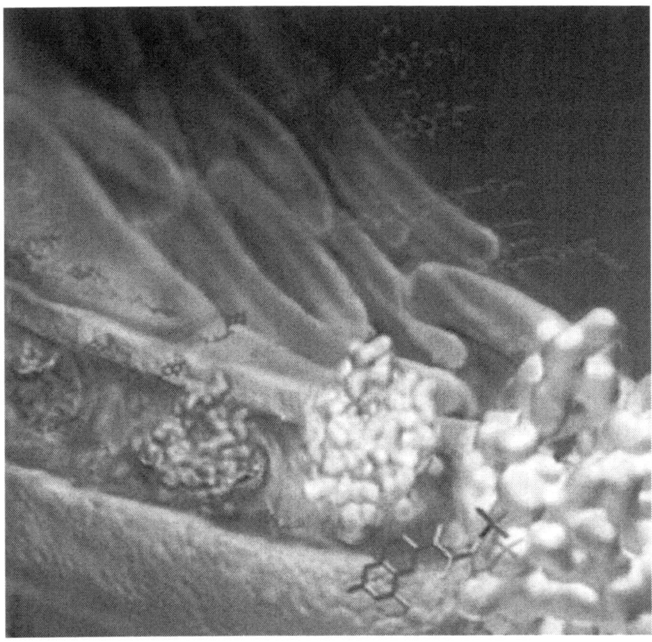

Fig. 11.5 : Microwave assisted synthesis of NP

The method offers an excellent reproducibility and avoids all the tedious process involved with synthesis of nanoparticles using micro-organisms.

In this research work, twenty five different anti-malignant plants are screened, including the popular anti-malignant Vinca (Vinca rosea) plant. Guava (Psidium guajava) leaf, which is also known for its anti-malignant effects, was selected after confirming its potentiality for further preparation of extract, synthesis of AuNP, and for screening. Studies shows that the bio-functionalized gold nanoparticles colloidal solution obtained as above will maintain its stability even after 30 weeks of storage.

The rate of synthesis of AuNP from microwave-exposed guava leaf extract is compared to those of chemical and physical methods, and it is noted that this has an excellent time-related advantage over benign and conventional method of metal nanoparticle synthesis from microorganisms and other plants.

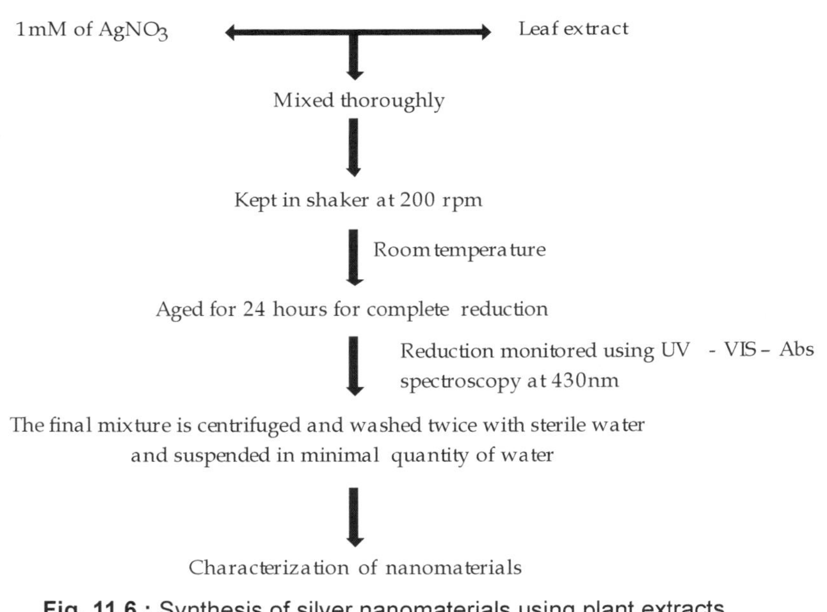

Fig. 11.6 : Synthesis of silver nanomaterials using plant extracts

---

10g of Fusarium oxysporum biomass has to be taken in a conica flask containing de ionized water

AgNo3 solution of (10-3M) has to be added, the reaction has to be carried out in risk

Periodically aliquots of the reaction can be removed

Fig. 11.7 : Protocol in Fungus

❑❑❑

Chapter-12

Synthesis of Core / Shell Nanoparticles

The most commonly used technique has been discussed and we have tried to generalize the type of particle that can be synthesized by each of these methods. Though it is not a rule that is applicable to all cases but it can be applied to most cases.

Polymerization

Radical Polymerization

The polymerization could be a free radical polymerization or an atom transfer radical polymerization. The process of atom transfer radical polymerization (ATRP) is better than the free radical polymerization as control of molecular weight and size of the particle

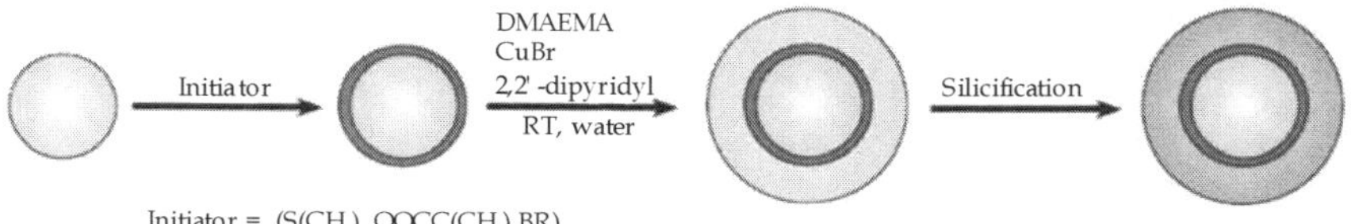

Fig. 12.1 : Schematic Representation of Procedure (Reprinted with permission from Kang et al. Copyright 2006. Institute of Physics Publishing.)

can be achieved.The coating of polymer on silica nanoparticles is generally done by ATRP. The surface of the silica particle is modified with a suitable initiator.

One of the methods would be to attach a bromine group to the surface of silica and add to solution containing the monomer of the shell polymer. The polymerization occurs and is depicted by change in the optical clarity of the solution. This method of attachment of bromine group to silica and forming polymer of t-butyl acrylate has been discussed by lei *et al.* Also Kang *et al.* discuss the formation of silica coated gold nanoparticles by a biomimetic approach through ATRP.

Chemical Oxidation Polymerization

The nanoparticle, if it has suitable groups attached to it allows the monomer which are generally aromatic compounds to form adduct. The monomer can be polymerized by adding suitable oxidizing agents. Generally metallic particles are given a coating of poly aromatic compounds by this method. Most metallic nanoparticles are formed by chemical reaction and then reduction. This implies there might be acidic or basic groups attached to the surface that induces modification. Such a method has been discussed by Jing *et al.* in silver/poly aniline and silver/poly pyrrole and PbS/poly pyrrole core/shell nanocomposite synthesis.

Sol Gel Method

This method of synthesis is used generally for the synthesis of metal/polymer core metal oxide shell nanoparticles in inorganic matrix that forms a gel, like silica. But some semiconductor nanoparticles have also been synthesized by this method. The steps can be stated in general as formation of the solution containing the salt of the metal and silica based compound. The solution is then heat treated and upon gelation, the metal salt is reduced in hydrogen atmosphere to metal nanoparticles. These are then subjected to heat in ordinary atmosphere forming an oxide shell on top of the metal nanoparticles. For polymer core and metal oxide shell the polymer nanoparticles can be added to metal salt solution and then oxidized. Iron/iron oxide,tin/ tinoxide,copper/ copper oxide in silica matrix have been synthesized by this method. There is some reference to synthesis of polypyyrole/iron oxide nanoparticles and Cdse/CdS nanoparticles by this method.

Reverse Micelle Method

One of the major concerns in the synthesis of nanoparticles and in specific core/shell nanoparticles is the achievement of control over size and morphology. This can be obtained by conducting the synthesis in emulsions or in solutions that form micelles. Micelles are formed by mixing aqueous reactant with suitable surfactant. Micelles act as the center for nucleation and epitaxial growth of nanoparticles. The molar ratio of the surfactant to water is the parameter that affects size and morphology of the resultant particles. These particles can be further processed to obtain core/shell structure by oxidation polymerization as discussed above. Else the surface can be coated with other metals or silica. Carpenter *et al.* have discussed reverse micelle technique for the synthesis of iron nanoparticles with a gold shell. Similar method has been proposed for cobaltplatinum core/shell structure by Kumbhar *et al.*.Wang *et al.* discusses iron oxide with gold core but has used ligand exchange reactivity to assemble particles into thin films. Polymeric particles can also be synthesized if the emulsion of the monomer is thermodynamically stable.

Mechanochemical Synthesis

The above discussed sub-topics themselves give an idea about the process to the readers. This sub-topic in itself houses varied synthesis techniques of which only two commonly used ones will be touched upon here. Mechanochemical synthesis as the name suggests involves mechanical and chemical means of nanoparticle synthesis.

Sonochemical Synthesis

The synthesis involves chemical reaction for nanoparticle synthesis and sonication to improve the speed of reaction, breakdown the particles and to enhance the dispersion of particles in the solvent. Ultrasonic irradiation of the frequency range 20 kHz to 1 MHz has been used in most of the sonication methods. Ultrasonic irradiation speeds up there action because of the localized cavities that are formed and they last only for a short time. Thus, these cavities act as micro reactors for the reaction to occur and the mechanical effects to also take place. The mixture of reactants in suitable solution is subjected to ultrasonic waves and the temperatureand pH maintained to obtain the nanoparticles dispersed in silica or in matrix material. The chemical reaction that occurs depends on the shell required as core is synthesized separately and added to the reactant mixture. Reduction reactions are carried out for metallic shells and *in situ* polymerization for polymeric shells/non-metallic shells. Composites such as iron oxide with gold shell and iron/cobalt alloy nanoparticles are ones with metallic shell and those such as silica/PAPBA, Ag2S/PVA, CuS/PVA are ones with nonmetallic shell that are synthesized by this method.

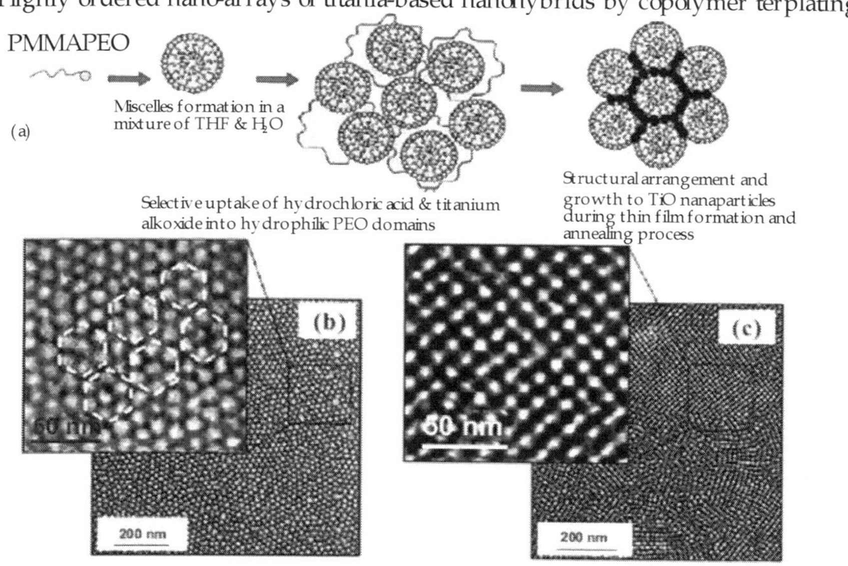

Fig. 12.2 : Assembly of Titania Hybrid and TEM Pictures (Reprinted with permission from J. Wang and All Yumono. Copyright 2006. Engineering research News NUS).

Electrodeposition

Formation of shell of nanoparticles with charged polymers or inorganic material can be carried out by this method. The electric pulse is varied like a square wave in general and it is found that the metal deposits during negative cycle while polymer deposits during positive cycle. This gives control over the size of the nanoparticles. The mode of operation is galvanization hence, the deposition occurs on one of the electrodes. The matrix material for nanoparticles can be the electrode or the electrolytic medium. Banerjee *et al.* detail on the synthesis of iron oxide shell iron in silica nanoparticles. Chipara *et al.* describe the synthesis of polypyyrole-iron nanoparticles.

Both the procedures are almost similar, the only difference being that in the iron oxide shell iron in silica nanoparticles, the electrolytic medium, silica gel, is the matrix for the nanocomposite while for the PPy-Fe,it doesn't have a matrix material and belongs to the class of nano-nanoparticles. The PPy-Fe can be dispersed in the desired matrix material later [30].Some other methods of synthesis include mechanical attrition, colloidal chemical synthesis, layer deposition and the like which are used widely for synthesizing magnetic nanoparticles. Some copper oxide nanoparticles are synthesized by reduction and pulverizing.

Synthesis of PLGA-PEG Block Copolymers

One useful strategy for modifying the physicochemical and biological properties of hydrophobic and biodegradable PLA, PGA, and PLGA has been to incorporate hydrophilic PEG segments. It is known that low molecular weight PEGs are easily excreted in humans. Many synthetic methods were developed to prepare various kinds of block copolymers with different block structures and compositions.

The biodegradation rate and hydrophilicity of block copolymers can be modulated by adjusting the ratio of its hydrophilic and hydrophobic constituents. Usually, PLGA-PEG block copolymers have shown quite different properties when compared to each constituting polymer. For this reason, PLGA-PEG block copolymers became a new family of biomaterials with their own unique properties, such as microphase separation, crystallinity, water-solubility, and biodegradability. Various kinds of block copolymers have been developed to date and can be classified according to their block structure as AB diblock, ABA, or BAB triblock, multi-block, branched block, star-shaped block, and graft block copolymers, in which A is a hydrophobic block made up of biodegradable polyesters and B is a hydrophilic PEG block, as shown in Figure.

Homo- and copolymers of LA and GA are usually synthesized by ring-opening polymerization of cyclic monomers. The block copolymers can be synthesized using various kinds of different catalysts, but also in the absence of catalysts. One of the most widely used catalysts is stannous octoate. Figure 3 shows a typical example for synthesis of PLGA-PEG block copolymers using stannous octoate. The terminal hydroxyl groups of PEG have been used as the initiating groups to synthesize block copolymers. Therefore, ring opening polymerization

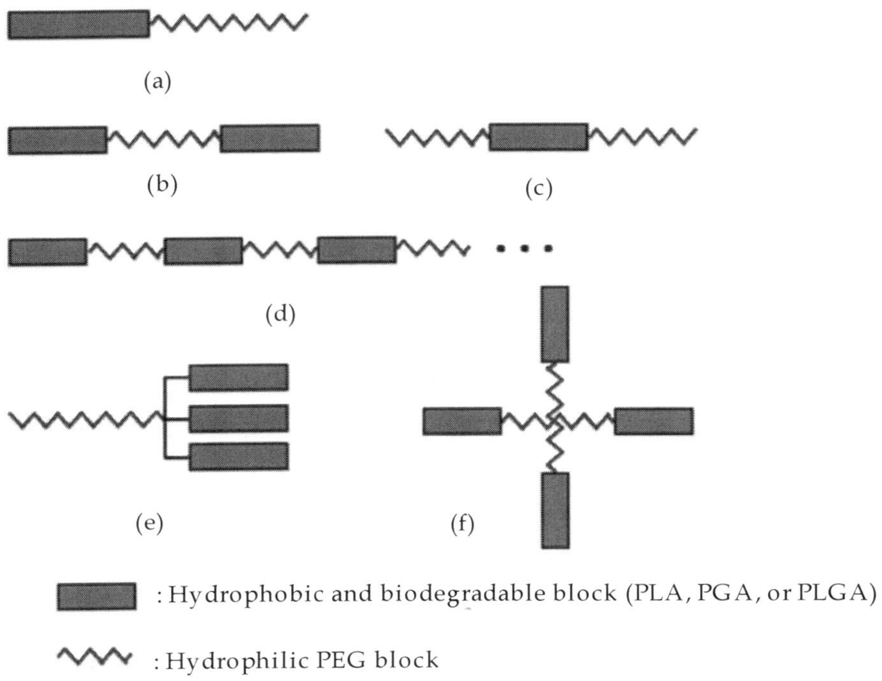

Fig. 12.3 : Schematic presentation of block copolymer structures : (a) A-B diblock (b) A-B-A (c) B-A-B (d) alternating multiblock (e) multi-armed structure and (f) star-shaped block.

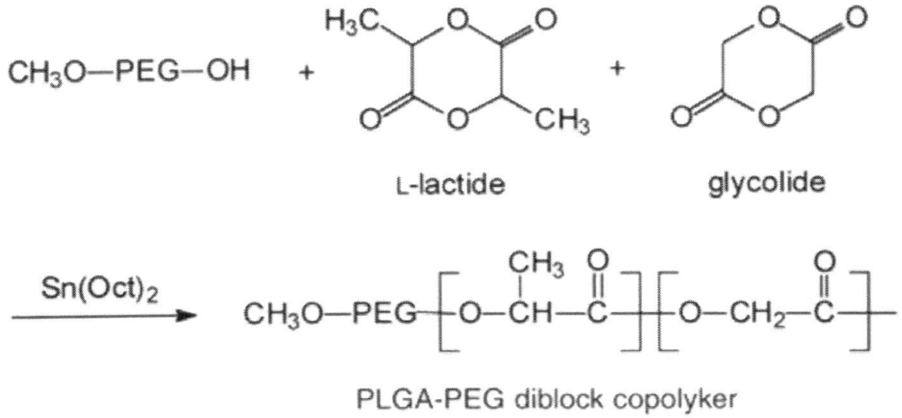

CH_3O—PEG—OH + L-lactide + glycolide $\xrightarrow{Sn(Oct)_2}$ CH_3O—PEG—[O—CH(CH_3)—C(=O)]—[O—CH_2—C(=O)]—

PLGA-PEG diblock copolyker

Fig. 12.4 : A typical synthetic route of a PLGA-PEG diblock copoymer.

of lactide and glycolide initiated by dihydroxy PEG or monomethoxy PEG can lead to A-B-A or A-B type block copolymers, respectively. B-A-B type block copolymers can be obtained by coupling the diblock copolymers using hexamethylene diisocyanate. Alternating multiblock copolymers of PLA and PEG can be synthesized by polycondensation reaction between dihydroxy PEG and dicarboxylated PLA. Dicarboxylated oligomeric PLAs were synthesized as macro-monomers by the condensation reaction of lactic acid in the presence of succinic acid. Kissel et al. reported star-shaped block copolymers from multi-arm PEG and lactide or lactide/glycolide.[15] Biodegradable star-shaped PLA-PEG and PLGA-PEG block copolymers can be synthesized by ring opening polymerization in the presence of 4- or 8-branched PEG using aluminum triethylene as a catalyst.

❑❑❑

Chapter-13

Synthesis of Carbon Nanotubes

Techniques have been developed to produce nanotubes in sizeable quantities, including arc discharge, laser ablation, high pressure carbon monoxide (HiPCO), and chemical vapor deposition (CVD). Most of these processes take place in vacuum or with process gases. CVD growth of CNTs can occur in vacuum or at atmospheric pressure.

Fig. 13.1 : Powder of carbon nanotubes

Large quantities of nanotubes can be synthesized by these methods; advances in catalysis and continuous growth processes are making CNTs more commercially viable.

Arc discharge

Nanotubes were observed in 1991 in the carbon soot of graphite electrodes during an arc discharge, by using a current of 100 amps that was intended to produce fullerenes. However the first macroscopic production of carbon nanotubes was made in 1992 by two researchers at NEC's Fundamental Research Laboratory. The method used was the same as in 1991. During this process, the carbon contained in the negative electrode sublimates because of the high discharge temperatures. Because nanotubes were initially discovered using this technique, it has been the most widely-used method of nanotube synthesis. The yield for this method is up to 30 per cent by weight and it produces both single- and multi-walled nanotubes with lengths of up to 50 micrometers with few structural defects.

Laser ablation

In the laser ablation process, a pulsed laser vaporizes a graphite target in a high-temperature reactor while an inert gas is bled into the chamber. Nanotubes develop on the cooler surfaces of the reactor as the vaporized carbon condenses. A water-cooled surface may be included in the system to collect the nanotubes.

This process was developed by Dr. Richard Smalley and co-workers at Rice University, who at the time of the discovery of carbon nanotubes, were blasting metals with a laser to produce various metal molecules. When they heard of the existence of nanotubes they replaced the metals with graphite to create multi-walled carbon nanotubes. Later that year the team used a composite of graphite and metal catalyst particles (the best yield was from a cobalt and nickel mixture) to synthesize single-walled carbon nanotubes.

The laser ablation method yields around 70% and produces primarily single-walled carbon nanotubes with a controllable diameter determined by the reaction temperature. However, it is more expensive than either arc discharge or chemical vapor deposition.

Chemical vapor deposition (CVD)

The catalytic vapor phase deposition of carbon was first reported in 1959, but it was not until 1993 that carbon nanotubes were formed by this process. In 2007, researchers at the University of Cincinnati (UC) developed a process to grow aligned carbon nanotube arrays of 18 mm length on a FirstNano ET3000 carbon nanotube growth system.

Fig. 13.2 : Nanotubes being grown by enhanced chemical vapor deposition

During CVD, a substrate is prepared with a layer of metal catalyst particles, most commonly nickel, cobalt, iron, or a combination. The metal nanoparticles can also be produced by other ways, including reduction of oxides or oxides solid solutions. The diameters of the nanotubes that are to be grown are related to the size of the metal particles. This can be controlled by patterned (or masked) deposition of the metal, annealing, or by plasma etching of a metal layer. The substrate is heated to approximately 700°C. To initiate the growth of nanotubes, two gases are bled into the reactor: a process gas (such as ammonia, nitrogen or hydrogen) and a carbon-containing gas (such as acetylene, ethylene, ethanol or methane). Nanotubes grow at the sites of the metal catalyst; the carbon-containing gas is broken apart at the surface of the catalyst particle, and the carbon is transported to the edges of the particle, where it forms the nanotubes. This mechanism is still being studied.

The catalyst particles can stay at the tips of the growing nanotube during the growth process, or remain at the nanotube base, depending on the adhesion between the catalyst particle and the substrate.

CVD is a common method for the commercial production of carbon nanotubes. For this purpose, the metal nanoparticles are mixed with a catalyst support such as MgO or Al_2O_3 to increase the surface area for higher yield of the catalytic reaction of the carbon feedstock with the metal particles. One issue in this synthesis route is the removal of the catalyst support via an acid treatment, which sometimes could destroy the original structure of the carbon nanotubes. However, alternative catalyst supports that are soluble in water have proven effective for nanotube growth.

If a plasma is generated by the application of a strong electric field during the growth process (plasma enhanced chemical vapor deposition*), then the nanotube growth will follow the direction of the electric field. By adjusting the geometry of the reactor it is possible to synthesize vertically aligned carbon nanotubes (i.e., perpendicular to the substrate), a morphology that has been of interest to researchers interested in the electron emission from nanotubes. Without the plasma, the resulting nanotubes are often randomly oriented. Under certain reaction conditions, even in the absence of a plasma, closely spaced nanotubes will maintain a vertical growth direction resulting in a dense array of tubes resembling a carpet or forest.

Of the various means for nanotube synthesis, CVD shows the most promise for industrial-scale deposition, because of its price/unit ratio, and because CVD is capable of growing nanotubes directly on a desired substrate, whereas the nanotubes must be collected in the other growth techniques. The growth sites are controllable by careful deposition of the catalyst. In 2007, a team from Meijo University demonstrated a high-efficiency CVD technique for growing carbon nanotubes from camphor. Researchers at Rice University, until recently led by the late Dr. Richard Smalley, have concentrated upon finding methods to produce large, pure amounts of particular types of nanotubes. Their approach grows long fibers from many small seeds cut from a single nanotube; all of the resulting fibers were found to be of the same diameter as the original nanotube and are expected to be of the same type as the original nanotube. Further characterization of the resulting nanotubes and improvements in yield and length of grown tubes are needed.

Natural, incidental, and controlled flame environments

Fullerenes and carbon nanotubes are not necessarily products of high-tech laboratories; they are commonly formed in such mundane places as ordinary flames, produced by burning methane, ethylene, and benzene, and they have been found in soot from both indoor and outdoor air. However, these naturally occurring varieties can be highly irregular in size and quality because the environment in which they are produced is often highly uncontrolled. Thus, although they can be used in some applications, they can lack in the high degree of uniformity necessary to meet many needs of both research and industry. Recent efforts have focused on producing more uniform carbon nanotubes in controlled flame environments. Nano-C, Inc of Westwood, Massachusetts, is producing flame synthesized single-walled carbon nanotubes. This method has promise for large-scale, low-cost nanotube synthesis, though it must compete with rapidly developing large scale CVD production.

❑❑❑

Chapter-14

Properties of Nanomaterials

1. Mechanical Properties of Nanomaterials

Scientific challenges in nanoscience and nanotechnology include the development of nanomaterials with novel mechanical properties.The need for scratch, mar and/or abrasion resistance is well established in various markets, including fingernail polishes, flooring, plastic glazing, headlamp covers and other automotive parts, transportation windows and optical lenses, where clear scratch-resistant coatings are used. Because the nanosize, many of their mechanical properties of the materials is modified, among others, hardness and elastic modulus, fracture toughness, scratch resistance, fatigue strength, and hardness. Energy dissipation, mechanical coupling within arrays of components, and mechanical nonlinearities are influenced by structuring components at the nanometer scale. This includes also the interpretation of unusual mechanical behavior (e.g., strengths approaching the theoretical limit) and the exploration of new ways to integrate diverse classes of mechanically functional materials on the nano-size.

Applications of Mechanical Properties of Nanomaterials

Tougher and harder cutting tools

Cutting tools made of nanomaterials, such as tungsten carbide, tantalum carbide, and titanium carbide, are much harder, much more wear-resistant, erosion-resistant, and last longer than their conventional (large-grained) counterparts. Also, for the miniaturization of microelectronic circuits, the industry needs micro drills (drill bits with diameter less than the thickness of an average human hair or 100 μm) with enhanced edge retention and far better wear resistance.Since nanocrystalline carbides are much stronger, harder, and wear-resistant, they are currently being used in these micro drills.

Automobiles with greater fuel efficiency

In automobiles, since nanomaterials are stronger, harder, and much more wear-resistant and erosion-resistant, they are envisioned to be used in spark plugs. Also, automobiles waste significant amounts of energy by losing the thermal energy generated by the engine. So, the engine cylinders are envisioned to be coated with nanocrystalline ceramics, such as zirconia and alumina, which retain heat much more efficiently that result in complete and efficient combustion of the fuel.

Aerospace components with enhanced performance characteristics

One of the key properties required of the aircraft components is the fatigue strength, which decreases with the component's age. The fatigue strength increases with a reduction in the grain size of the material. Nanomaterials provide such a significant reduction in the grain size over conventional materials that the fatigue life is increased by an average of 200-300%.In spacecrafts, elevated-temperature strength of the material is crucial because the components (such as rocket engines, thrusters, and vectoring nozzles) operate at much higher temperatures than aircrafts and higher speeds. Nanomaterials are perfect candidates for spacecraft applications, as well.

Ductile ceramics

Ceramics are very hard, brittle, and hard to machine even at high temperatures.However, with a reduction in grain size, their properties change drastically. Nanocrystalline ceramics can be pressed and sintered into various shapes at significantly lower temperatures. Zirconia, for example, is a hard, brittle ceramic, has even been rendered superplastic, i. e., it can deformed to great lengths (up to 300% of its original length). However, these ceramics must possess nanocrystalline grains to be superplastic. Ceramics based on silicon nitride (Si3N4) and silicon carbide (SiC), have been used in automotive applications as high-strength springs, ball bearings, and valve lifters, and because they possess good formability and machinabilty combined with excellent physical, chemical, and mechanical properties. They are also used as components in high-temperature furnaces.

Better insulation materials

Aerogels are nanocrystalline porous and extremely lightweight materials and can withstand 100 times their weight.They are currently being used for insulation in offices, homes, etc. They are also being used as materials for "smart" windows, which darken when the sun is too bright and they lighten themselves otherwise.

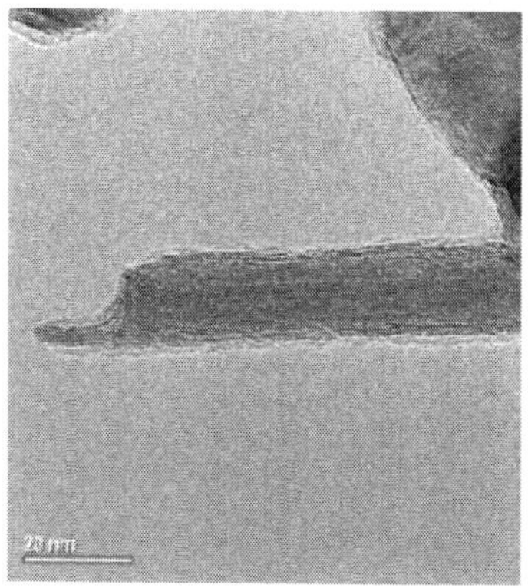

Fig. 14.1 : Making a Nanowire

2. Optical Properties of Nanomaterials

Nanocrystalline systems have attracted much interest for their novel optical properties, which differ remarkably from bulk crystals. Key contributory factors include quantum confinement of electrical carriers within nanoparticles, efficient energy and charge transfer over nanoscale distances and in many systems a highly enhanced role of interfaces. With the growing technology of these materials, it is increasingly necessary to understand the detailed basis for nanophotonic properties.The linear and nonlinear optical properties of such materials can be finely tailored by controlling the crystal dimensions, and the chemistry of their surfaces, fabrication technology becomes a key factor for the applications. Surface Plasmons (SP) are the origin of the color of nanomaterials.An SP is a natural oscillation of the electron gas inside a given nanosphere. If the sphere is small compared to a wavelength of light, and the light has a frequency close to that of the SP, then the SP will absorb energy. The frequency of the SP depends on the dielectric function of the nanomaterial, and the shape of the nanoparticle. For a gold spherical particle, the frequency is about 0.58 of the bulk plasma frequency. Thus, although the bulk plasma frequency is in the UV, the SP frequency is in the visible (close to 520 nm)Suppose we have a suspension of nanoparticles in a host. If a wave of light is applied, the local electric field may be hugely enhanced near an SP resonance. If so, one expects various nonlinear susceptibilities, which depend on higher powers of the electric field, to be enhanced even more.

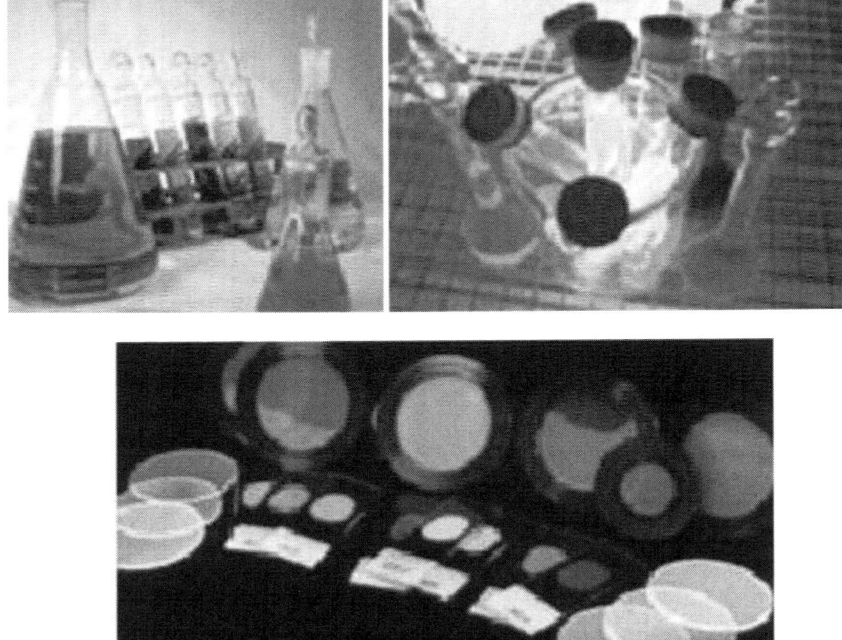

Fig. 14.2 : Depending on the particle's size, different colors are seen. In these images we see the different color produced by gold particles of different sizes.

Fig. 14.3 : Stained glass window at Church of the Immaculate Conception in Pawhuska, Oklahoma. The ruby color is due to embedded gold nanoparticles of different sizes.

Applications of Optical Properties of Nanomaterials

Glues containing nanoparticles have optical properties that give rise to uses in optoelectronics.Casings, containing nanoparticles used in electronic devices, such as computers, offer improved shielding against electromagnetic interference. Electrochromic, devices are similar to liquid-crystal displays (LCD), are been developed with nanomaterials. The incorporation of nanomaterials in surface coatings can provide long-term abrasion resistance without significantly effecting optical clarity, gloss, color or physical properties.

3. Magnetic Properties of Nanomaterials

The strength of a magnet is measured in terms of coercivity and saturation magnetization values. These values increase with a decrease in the grain size and an increase in the specific surface area (surface area per unit volume) of the grains. Therefore nanomaterials present also good properties in this field.

Applications of Magnetic Properties of Nanomaterials

High-power magnets

Magnets made of nanocrystalline yttrium-samarium-cobalt grains possess very unusual magnetic properties due to their extremely large surface area. Typical applications for these high-power rare-earth magnets include quieter submarines, automobile alternators, land-based power generators, and motors for ships, ultra-sensitive analytical instruments, and magnetic resonance imaging (MRI) in medical diagnostics.

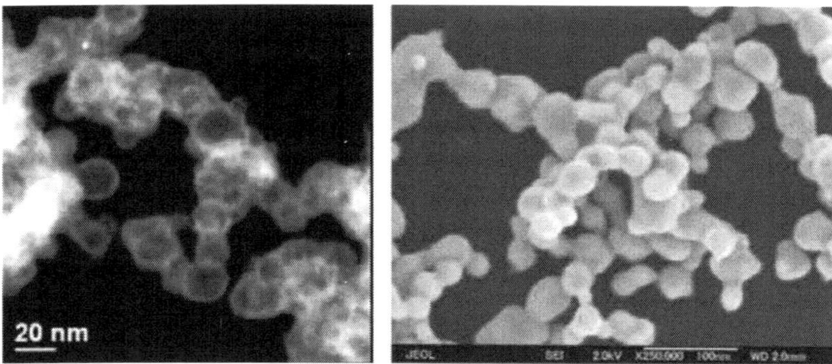

Fig. 14.4 : Silver nanoparticles covered with carbon film.

4. Electrical Properties of Nanomaterials

Nanomaterials can hold considerably more energy than conventional because of their large grain boundary (surface) area. They are materials in which an optical absorption band can be introduced, or an existing band can be altered by the passage of current through these materials, or by the application of an electric field.

Applications of Electrical Properties of Nanomaterials

High energy density batteries Conventional and rechargeable batteries are used in almost all applications that require electric power. The energy density (storage capacity) of these batteries is quite low requiring frequent recharging. Nanocrystalline materials are good candidates for separator plates in batteries because they can hold considerably more energy than conventional ones. Nickel-metal hydride batteries made of nanocrystalline nickel and metal hydrides are envisioned to require far less frequent recharging and to last much longer.

Large electrochromic display devices

An electrochromic device consists of materials in which an optical absorption band can be introduced, or an existing band can be altered by the passage of current through the materials, or by the application of an electric field. They are similar to liquid-crystal displays (LCD) commonly used in calculators and watches and are primarily used in public billboards and ticker boards to convey information. The resolution, brightness, and contrast of these devices depend on the tungstic acid gel's grain size. Hence, nanomaterials, such as tungstic oxide gel, are being explored for this purpose.

5. Chemical Properties of Nanomaterials

One of the important factors for the chemical applications of nanomaterials is the increment of their surface area which increases the chemical activity of the material.

Applications of Chemical Properties of Nanomaterials

Due to their enhanced chemical activity, nanostructural materials can be used as catalysts to react with such noxious and toxic gases as carbon monoxide and nitrogen oxide in automobile catalytic converters and power generation equipment to prevent environmental pollution arising from burning gasoline and coal. Fuel cell technology is another important application of the noble metal nanoparticles relating the catalysis of the reactions.In the present, the fuel cell catalysts are based on platinum group metals (PGM). Pt and Pt-Ru alloys are some of the most frequently used catalysts from this group.In fact, the use of these metals is one major factor for cell costs, which has been one of the major drawbacks preventing it from growing into a more important technology. One possibility to produce economical catalysts is the use of bimetallic nanoparticles.

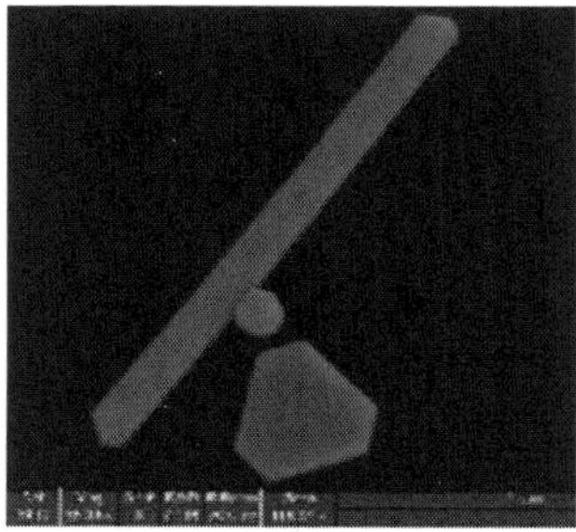

Fig. 14.5 : Gold in Different Shapes

❑❑❑

Chapter-15

Properties of Carbon Nanotubes and Block Copolymers

Strength

Carbon nanotubes are the strongest and stiffest materials yet discovered in terms of tensile strength and elastic modulus respectively. This strength results from the covalent sp^2 bonds formed between the individual carbon atoms. In 2000, a multi-walled carbon nanotube was tested to have a tensile strength of 63 gigapascals (GPa). (This, for illustration, translates into the ability to endure tension of 6300 kg on a cable with cross-section of 1 mm^2.) Since carbon nanotubes have a low density for a solid of 1.3 to 1.4 $g \cdot cm^{-3}$, its specific strength of up to 48,000 $kN \cdot m \cdot kg^{-1}$ is the best of known materials, compared to high-carbon steel's 154 $kN \cdot m \cdot kg^{-1}$. Under excessive tensile strain, the tubes will undergo plastic deformation, which means the deformation is permanent. This deformation begins at strains of approximately 5% and can increase the maximum strain the tubes undergo before fracture by releasing strain energy.

CNTs are not nearly as strong under compression. Because of their hollow structure and high aspect ratio, they tend to undergo buckling when placed under compressive, torsional or bending stress.

The above discussion referred to axial properties of the nanotube, whereas simple geometrical considerations suggest that carbon nanotubes should be much softer in the radial direction than along the tube axis. Indeed, TEM observation of radial elasticity suggested that even the van der Waals forces can deform two adjacent nanotubes. Nanoindentation experiments, performed by several groups on multiwalled carbon nanotubes, indicated Young's modulus of the order of several GPa confirming that CNTs are indeed rather soft in the radial direction.

Kinetic

Multi-walled nanotubes exhibit a striking telescoping property whereby an inner nanotube core may slide, almost without friction, within its outer nanotube shell thus creating an atomically perfect linear or rotational bearing. This is one of the first true examples of molecular nanotechnology, the precise positioning of atoms to create useful machines. Already this property has been utilized to create the world's smallest rotational motor.

Electrical

Because of the symmetry and unique electronic structure of graphene, the structure of a nanotube strongly affects its electrical properties. For a given (n,m) nanotube, if $n = m$, the nanotube is metallic; if n “ m is a multiple of 3, then the nanotube is semiconducting with a very small band gap, otherwise the nanotube is a moderate semiconductor. Thus all armchair ($n = m$) nanotubes are metallic, and nanotubes (5,0), (6,4), (9,1), etc. are semiconducting. In theory, metallic nanotubes can carry an electrical current density of 4×10^9 A/cm^2 which is more than 1,000 times greater than metals such as copper.

Thermal

All nanotubes are expected to be very good thermal conductors along the tube, exhibiting a property known as "ballistic conduction," but good insulators laterally to the tube axis. It is predicted that carbon nanotubes will be able to transmit up to 6000 W ·m$^{″1}$ ·K$^{″1}$ at room temperature; compare this to copper, a metal well-known for its good thermal conductivity, which transmits 385 W ·m$^{″1}$ ·K$^{″1}$. The temperature stability of carbon nanotubes is estimated to be up to 2800 °C in vacuum and about 750 °C in air.

Defects

As with any material, the existence of a crystallographic defect affects the material properties. Defects can occur in the form of atomic vacancies. High levels of such defects can lower the tensile strength by up to 85%. Another form of carbon nanotube defect is the Stone Wales defect, which creates a pentagon and heptagon pair by rearrangement of the bonds. Because of the very small structure of CNTs, the tensile strength of the tube is dependent on its weakest segment in a similar manner to a chain, where the strength of the weakest link becomes the maximum strength of the chain.

Crystallographic defects also affect the tube's electrical properties. A common result is lowered conductivity through the defective region of the tube. A defect in armchair-type tubes (which can conduct electricity) can cause the surrounding region to become semiconducting, and single monoatomic vacancies induce magnetic properties.

Crystallographic defects strongly affect the tube's thermal properties. Such defects lead to phonon scattering, which in turn increases the relaxation rate of the phonons. This reduces the mean free path and reduces the thermal conductivity of nanotube structures. Phonon transport simulations indicate that substitutional defects such as nitrogen or boron will primarily lead to scattering of high-frequency optical phonons. However, larger-scale defects such as Stone Wales defects cause phonon scattering over a wide range of frequencies, leading to a greater reduction in thermal conductivity.

One-dimensional transport

Due to their nanoscale dimensions, electron transport in carbon nanotubes will take place through quantum effects and will only propagate along the axis of the tube. Because of this special transport property, carbon nanotubes are frequently referred to as "one-dimensional" in scientific articles.

Toxicity

Determining the toxicity of carbon nanotubes has been one of the most pressing questions in nanotechnology. Unfortunately such research has only just begun and the data is still fragmentary and subject to criticism. Preliminary results highlight the difficulties in evaluating the toxicity of this heterogeneous material. Parameters such as structure, size distribution, surface area, surface chemistry, surface charge, and agglomeration state as well as purity of the samples, have considerable impact on the reactivity of carbon nanotubes. However, available data clearly show that, under some conditions, nanotubes can cross membrane barriers, which suggests that if raw materials reach the organs they can induce harmful effects such as inflammatory and fibrotic reactions. A study led by Alexandra Porter from the University of Cambridge shows that CNTs entered human cells and accumulate in the cytoplasm, causing cell death.

Results of rodent studies collectively show that regardless of the process by which CNTs were synthesized and the types and amounts of metals they contained, CNTs were capable of producing inflammation, epithelioid granulomas (microscopic nodules), fibrosis, and biochemical/toxicological changes in the lungs. Comparative toxicity studies in which mice were given equal weights of test materials showed that SWCNTs were more toxic than quartz, which is considered a serious occupational health hazard when chronically inhaled. As a control, ultrafine carbon black was shown to produce minimal lung responses.

The needle-like fiber shape of CNTs, similar to asbestos fibers, raises fears that widespread use of carbon nanotubes may lead to mesothelioma, cancer of the lining of the lungs often caused by exposure to asbestos. Although further research is required, current studies suggest that, under certain conditions, especially those involving chronic exposure, carbon nanotubes can pose a serious risk to human health.

Properties of PLGA-PEG Block Copolymers

One typical characteristic of PLGA degradation is autocatalysis by which heterogeneous bulk degradation is observed with a decrease in pH. Carboxylic end groups of degraded products, oligomeric PLGA, can accelerate the degradation and decrease the local pH in PLGA formulations. This locally acidified environment is known to be a main reason for protein inactivation and often requires incorporation of antacids, such as $Mg(OH)_2$, into the polymers for protein stabilization by neutralization.[16] In addition, other limitations, such as hydrophobicity, brittleness, and toxicity, have been reported with PLGA formulations. Block copolymers containing hydrophilic PEG segments have attracted considerable attention as an alternative approach for overcoming such undesirable effects and improving the properties in applications of PLGA as drug delivery vehicles. Through various synthetic processes, diverse block copolymers with a wide range of molecular weights, chemical structure, and hydrophilic/hydrophobic block ratios have been prepared and used in controlled drug delivery.

The chemical composition and molecular weight of block copolymers determines their water-solubility and degradation kinetics. Polymers with low molecular weight or composed of shorter hydrophobic blocks are soluble in water, whereas high molecular weight polymers and polymers with longer hydrophobic blocks are not soluble but swell in water. In general, the degradation time will be shorter for low molecular weight polymers, more hydrophilic polymers, more amorphous polymers, and copolymers with higher content of glycolide. Therefore, at identical conditions, low molecular weight copolymers of lactide and glycolide will degrade relatively rapidly, whereas the high molecular weight homopolymers, PLA, and PGA will degrade much more slowly.

❑❑❑

Chapter-16

Nanoparticle Characterization

Nanoparticle characterization is necessary to establish understanding and control of nanoparticle synthesis and applications. Characterization is done by using a variety of different techniques, mainly drawn from materials science. Common techniques are electron microscopy (TEM,SEM), atomic force microscopy (AFM), dynamic light scattering (DLS), x-ray photoelectron spectroscopy (XPS), powder X-ray diffraction (XRD), Fourier transform infrared spectroscopy (FTIR), matrix-assisted laser desorption/ionization time-of-flight mass spectrometry (MALDI-TOF),ultraviolet-visible spectroscopy and dual polarisation interferometry.

Whilst the theory has been known for over a century (see Robert Brown), the technology for Nanoparticle tracking analysis (NTA) allows direct tracking of the Brownian motion and this method therefore allows the sizing of individual nanoparticles in solution.

Particle size Characterization

Particle size is an important parameter for many industrial processes. The chemical, optical and mechanical properties, the mixing behavior and the bio-distribution of many raw materials and end-products are affected by the size and shape of the particles they consist of. Therefore, monitoring particle dimensions is an important step in product optimization.

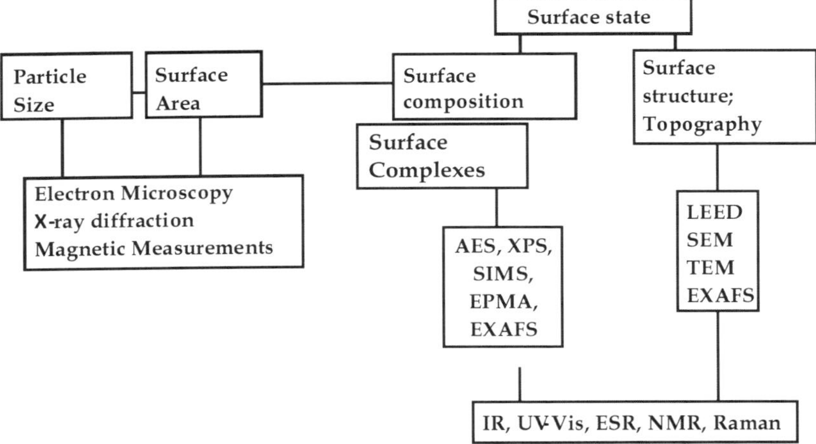

Fig. 16.1 : Common Methods of Nanoparticle Characterization

Nanoparticle Size Measurement Techniques

1. *Imaging methods:*

   Scanning Electron Microscope (SEM), Transmission Electron Microscope (TEM), Atomic Force Microscope (AFM) and etc.

2. *Light Scattering Methods:*

   Static Light Scattering (microparticles), Dynamic Light Scattering (nanoparticles)

3. *Separation Methods*

   Capillary Chromatography, Field Flow Fractionation combined with UV absorption methods

Techniques

Air-jet sieve analysis

Dry powders are placed on a sieve. This sieve is mounted in an airtight container. Air is blown upwards through the sieve from a rotating nozzle to fluidize the sample. The exit airflow carries undersized particles downward through the sieve to a collection canister. Starting with the finest sieve, the amount of material passing is determined by weighing. The retained sample is transferred to the next larger sieve size, and the procedure is repeated until sieving has been done on all required sieves in succession.

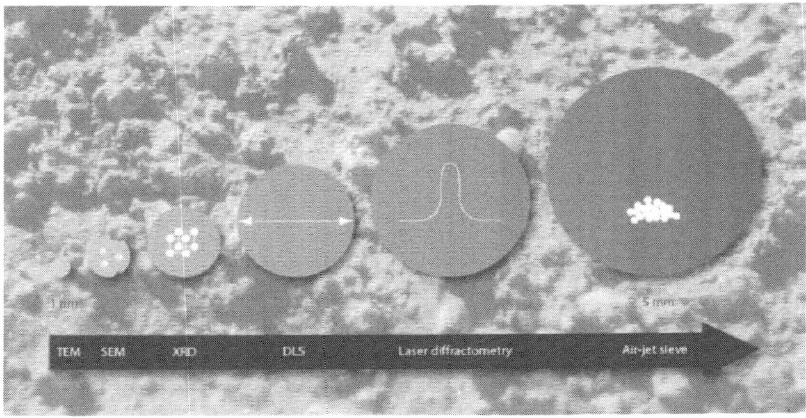

Fig. 16.2 : Size measurement techniques

Laser diffractometry

A suspension is pumped through a measuring cell and illuminated by a laser beam. When particles of different sizes pass a laser beam they cause the laser light to be scattered at angles that are inversely proportional to the particle size applicable to both suspensions and powders.

Dynamic Light Scattering (DLS)

A laser shines on a liquid containing particles (a suspension, emulsion, ...). The Brownian motion of the particles causes the intensity of the scattered light to vary in time. This fluctuation in intensity is related to the size of the particles; smaller particles move quicker than larger particles. Analyzing the intensity fluctuations yields diffusion coefficients and –by using the Stokes-Einstein relation – particle sizes.

X-ray Diffraction (XRD)

When X-rays are directed on a material of which the atoms are regularly arranged, the radiation is diffracted in specific directions. XRD uses this phenomenon to study the crystal structure and microstructure of crystalline solids. The limited sizes of crystallites in a powder or polycrystalline material broaden the peaks in the diffraction profile. This effect can be used to determine the size of small (<100 nm) crystallites, i.e. the size of single crystalline domains within a particle or bulk material.

Scanning Electron Microscopy (SEM)

SEM is an imaging and analysis technique based on the detection of electrons and X-rays that are emitted from a material when irradiated by a scanning electron beam. Both topographical and compositional information can be obtained with high spatial resolution.Imaging allows us to distinguish betweenprimary particle and agglomerate sizes. Dedicated software allows for automated image analysis, e.g. to determine particle size distributions.

Transmission Electron Microscopy (TEM)

TEM is a microscopy technique that allows for high resolution imaging as well as for chemical and structural studies on an atomic scale. An image is formed by electrons that have passed through the sample. Suspensions can either be studied after drying on a supporting film or can be quench-frozen, allowing for inspection of particle properties in the aqueous state.

Table 16.1 : Comparison of different techniques

	Sample type	Information obtained	Sizes
Air jet sleve	Powders	Weight fraction distribution	20-5600 μm
Laser diffractometry	Powders Suspensions	Size distribution	0.02-2000 μm
DLS	Suspensions Emulsions	Size distribution	1-1000 nm
XRD	Powders Polycrystaline material	Average grain size Crystal structure	2-100 nm
SEM	Powders Suspensions	Size distribution Shape Surface morphology	10 nm-mm's
TEM	Powders Suspensions Emulsions	Size distribution Shape Crystal structure Core/shell structure	1-400 nm

❑❑❑

Chapter-17

X-Ray and Light Scattering Technique

X-ray scattering techniques are a family of non-destructive analytical techniques which reveal information about the crystallographic structure, chemical composition, and physical properties of materials and thin films. These techniques are based on observing the scattered intensity of an X-ray beam hitting a sample as a function of incident and scattered angle, polarization, and wavelength or energy.

X-ray diffraction techniques

X-ray diffraction yields the atomic structure of materials and is based on the elastic scattering of X-rays from the electron clouds of the individual atoms in the system. The most comprehensive description of scattering from crystals is given by the dynamical theory of diffraction.

- Single-crystal X-ray diffraction is a technique used to solve the complete structure of crystalline materials, ranging from simple inorganic solids to complex macromolecules, such as proteins.
- Powder diffraction (XRD) is a technique used to characterise the crystallographic structure, crystallite size (grain size), and

preferred orientation in polycrystalline or powdered solid samples. Powder diffraction is commonly used to identify unknown substances, by comparing diffraction data against a database maintained by the International Centre for Diffraction Data. It may also be used to characterize heterogeneous solid mixtures to determine relative abundance of crystalline compounds and, when coupled with lattice refinement techniques, such as Rietveld refinement, can provide structural information on unknown materials. Powder diffraction is also a common method for determining strains in crystalline materials. An effect of the finite crystallite sizes is seen as a broadening of the peaks in an X-ray diffraction as is explained by the Scherrer Equation.

- ❑ Thin film diffraction and grazing incidence X-ray diffraction may be used to characterize the crystallographic structure and preferred orientation of substrate-anchored thin films.
- ❑ High-resolution X-ray diffraction is used to characterize thickness, crystallographic structure, and strain in thin epitaxial films. It employs parallel-beam optics.
- ❑ X-ray pole figure analysis enables one to analyze and determine the distribution of crystalline orientations within a crystalline thin-film sample.
- ❑ X-ray rocking curve analysis is used to quantify grain size and mosaic spread in crystalline materials.

Scattering techniques

Elastic scattering

Materials that do not have long range order may also be studied by scattering methods that rely on elastic scattering of monochromatic X-rays.

- ❑ Small angle X-ray scattering (SAXS) probes structure in the nanometer to micrometer range by measuring scattering intensity at scattering angles 2¸ close to 0°.[2]
- ❑ X-ray reflectivity is an analytical technique for determining thickness, roughness, and density of single layer and multilayer thin films.
- ❑ Wide angle X-ray scattering (WAXS), a technique concentrating on scattering angles 2¸ larger than 5°.

Inelastic scattering

When the energy and angle of the inelastically scattered X-rays are monitored scattering techniques can be used to probe the electronic band structure of materials.

- ❑ Compton scattering
- ❑ Resonant inelastic X-ray scattering (RIXS)
- ❑ X-ray Raman scattering

X-ray reflection in accordance with Bragg's Law

When a crystal is bombarded with X-rays of a fixed wavelength (similar to spacing of the atomic-scale crystal lattice planes) and at certain incident angles, intense reflected X-rays are produced when the wavelengths of the scattered X-rays interfere constructively. In order for the waves to interfere constructively, the differences in the travel path must be equal to integer multiples of the wavelength. When this constructive interference occurs, a diffracted beam of X-rays will leave the crystal at an angle equal to that of the incident beam.

To illustrate this feature, consider a crystal with crystal lattice planar distances d (right). Where the travel path length difference between the ray paths ABC and A'B'C' is an integer multiple of the wavelength, constructive interference will occur for a combination of that specific wavelength, crystal lattice planar spacing and angle of incidence (˜). Each rational plane of atoms in a crystal will undergo refraction at a single, unique angle (for X-rays of a fixed wavelength).

The general relationship between the wavelength of the incident X-rays, angle of incidence and spacing between the crystal lattice planes of atoms is known as Bragg's Law, expressed as:

$$n \lambda = 2d \sin \theta$$

where n (an integer) is the "order" of reflection, » is the wavelength of the incident X-rays, d is the interplanar spacing of the crystal and ˜ is the angle of incidence.

X-ray powder diffraction (XRD) is a rapid analytical technique primarily used for phase identification of a crystalline material and can provide information on unit cell dimensions. The analyzed material is finely ground, homogenized, and average bulk composition is determined

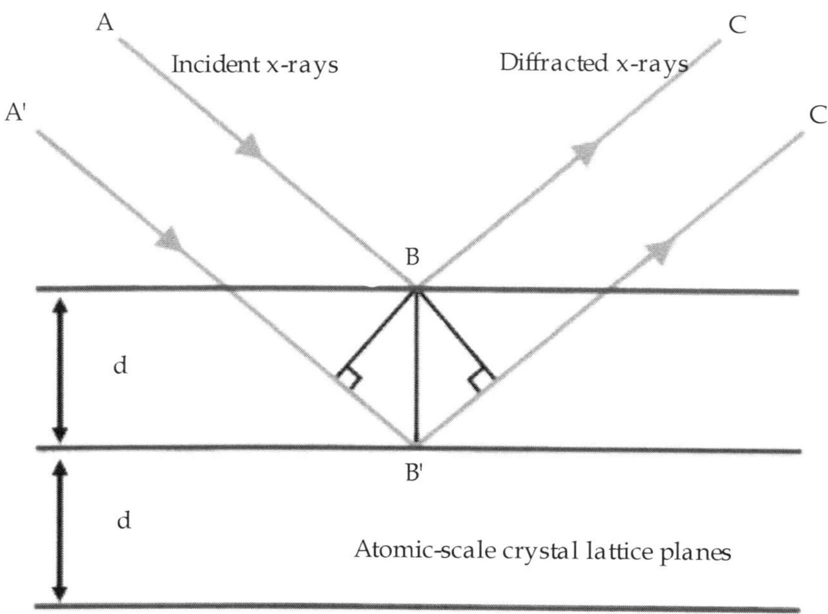

Fig. 17.1 : X-ray Powder Diffraction (XRD)

Fundamental Principles of X-ray Powder Diffraction (XRD)

Max von Laue, in 1912, discovered that crystalline substances act as three-dimensional diffraction gratings for X-ray wavelengths similar to the spacing of planes in a crystal lattice. X-ray diffraction is now a common technique for the study of crystal structures and atomic spacing.

X-ray diffraction is based on constructive interference of monochromatic X-rays and a crystalline sample. These X-rays are generated by a cathode ray tube, filtered to produce monochromatic radiation, collimated to concentrate, and directed toward the sample. The interaction of the incident rays with the sample produces constructive interference (and a diffracted ray) when conditions satisfy Bragg's Law ($n\lambda=2d \sin \theta$). This law relates the wavelength of electromagnetic radiation to the diffraction angle and the lattice spacing in a crystalline sample. These diffracted X-rays are then detected, processed and counted. By scanning the sample through a range of 2¸angles, all possible diffraction directions of the lattice should be attained due to the random orientation of the powdered material. Conversion of the diffraction peaks to d-spacings allows identification of the mineral because each mineral has a set of unique d-spacings. Typically, this is achieved by comparison of d-spacings with standard reference patterns.

All diffraction methods are based on generation of X-rays in an X-ray tube. These X-rays are directed at the sample, and the diffracted rays are collected. A key component of all diffraction is the angle between the incident and diffracted rays. Powder and single crystal diffraction vary in instrumentation beyond this.

X-ray Powder Diffraction (XRD) Instrumentation - How Does It Work?

X-ray diffractometers consist of three basic elements: an X-ray tube, a sample holder, and an X-ray detector.

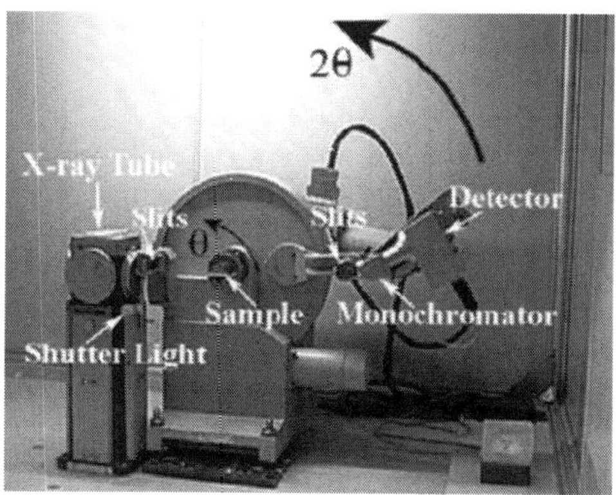

Fig. 17.2 : X-ray Diffractometer

X-rays are generated in a cathode ray tube by heating a filament to produce electrons, accelerating the electrons toward a target by applying a voltage, and bombarding the target material with electrons. When electrons have sufficient energy to dislodge inner shell electrons of the target material, characteristic X-ray spectra are produced. These spectra consist of several components, the most common being $K_{\pm}$ and K_{2}. $K_{\pm}$ consists, in part, of $K_{\pm1}$ and $K_{\pm2}$. $K_{\pm1}$ has a slightly shorter wavelength and twice the intensity as $K_{\pm2}$. The specific wavelengths are characteristic of the target material (Cu, Fe, Mo, Cr). Filtering, by foils or crystal monochrometers, is required to produce monochromatic X-rays needed for diffraction. $K_{\pm1}$and $K_{\pm2}$ are sufficiently close in wavelength such that a weighted average of the two is used. Copper is the most common target material for single-crystal diffraction, with CuK$_{\pm}$ radiation = 1.5418Å. These X-rays are collimated and directed onto the sample. As the sample and detector are rotated, the intensity of the reflected X-rays is recorded. When the geometry of the incident X-rays impinging the sample satisfies the Bragg Equation, constructive interference occurs and a peak in intensity occurs. A detector records and processes this X-ray signal and converts the signal to a count rate which is then output to a device such as a printer or computer monitor.

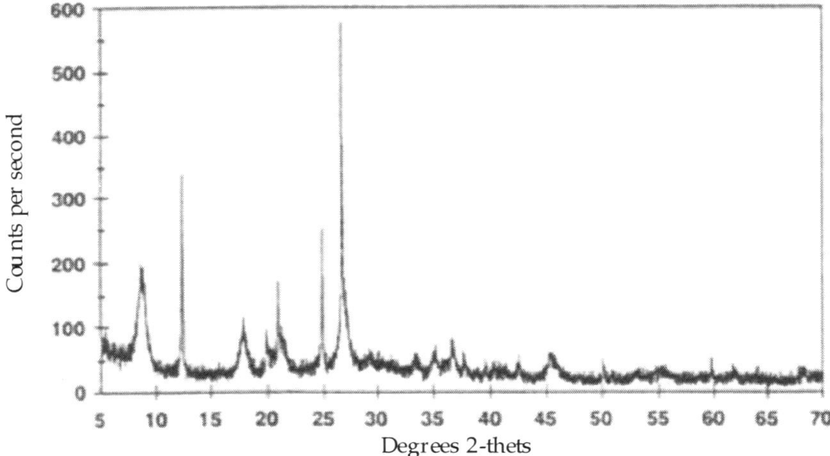

Fig. 17.3 : X-ray powder diffractogram.

Peak positions occur where the X-ray beam has been diffracted by the crystal lattice. The unique set of d-spacings derived from this patter can be used to 'fingerprint' the mineral.

The geometry of an X-ray diffractometer is such that the sample rotates in the path of the collimated X-ray beam at an angle ¸ while the X-ray detector is mounted on an arm to collect the diffracted X-rays and rotates at an angle of 2¸. The instrument used to maintain the angle and rotate the sample is termed a *goniometer*. For typical powder patterns, data is collected at 2¸ from ~5° to 70°, angles that are preset in the X-ray scan.

Applications

X-ray powder diffraction is most widely used for the identification of unknown crystalline materials (e.g. minerals, inorganic compounds).

Determination of unknown solids is critical to studies in geology, environmental science, material science, engineering and biology.

Other applications include:

- characterization of crystalline materials
- identification of fine-grained minerals such as clays and mixed layer clays that are difficult to determine optically
- determination of unit cell dimensions
- measurement of sample purity

With specialized techniques, XRD can be used to:

- determine crystal structures using Rietveld refinement
- determine of modal amounts of minerals (quantitative analysis)
- characterize thin films samples by:
  - Determining lattice mismatch between film and substrate and to inferring stress and strain
  - Determining dislocation density and quality of the film by rocking curve measurements
  - Measuring superlattices in multilayered epitaxial structures
  - Determining the thickness, roughness and density of the film using glancing incidence X-ray reflectivity measurements
- make textural measurements, such as the orientation of grains, in a polycrystalline sample

Strengths and Limitations of X-ray Powder Diffraction (XRD)?

Strengths

- Powerful and rapid (< 20 min) technique for identification of an unknown mineral
- In most cases, it provides an unambiguous mineral determination
- Minimal sample preparation is required
- XRD units are widely available
- Data interpretation is relatively straight forward

Limitations

- Homogeneous and single phase material is best for identification of an unknown
- Must have access to a standard reference file of inorganic compounds (d-spacings, *hkl*s)
- Requires tenths of a gram of material which must be ground into a powder
- For mixed materials, detection limit is ~ 2% of sample
- For unit cell determinations, indexing of patterns for non-isometric crystal systems is complicated
- Peak overlay may occur and worsens for high angle 'reflections'

Light scattering

Light scattering is a form of scattering in which light is the form of propagating energy which is scattered. Light scattering can be thought of as the deflection of a ray from a straight path, for example by irregularities in the propagation medium, particles, or in the interface between two media. Deviations from the law of reflection due to irregularities on a surface are also usually considered to be a form of scattering. When these irregularities are considered to be random and dense enough that their individual effects average out, this kind of scattered reflection is commonly referred to as diffuse reflection.

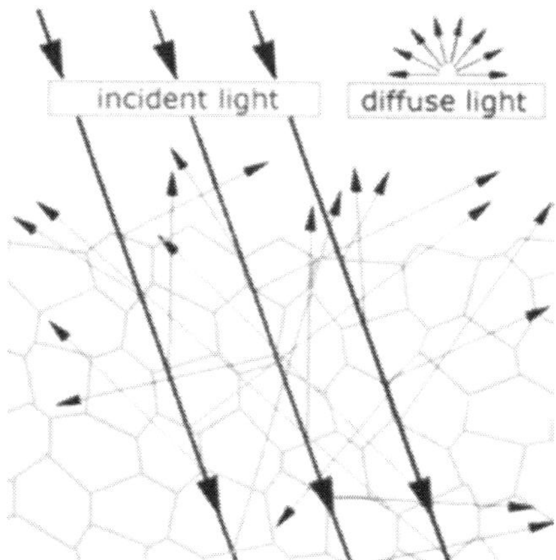

Fig. 17.4 : Light Scattering

Most objects that one sees are visible due to light scattering from their surfaces. Indeed, this is our primary mechanism of physical observation.[1][2] Scattering of light depends on the wavelength or frequency of the light being scattered. Since visible light has wavelength on the order of a micron, objects much smaller than this cannot be seen, even with the aid of a microscope. Colloidal particles as small as 1 μm have been observed directly in aqueous suspension.

Mechanisms of diffuse reflection include surface scattering from roughness and subsurface scattering from internal irregularities such as grain boundaries in polycrystalline solids.

The transmission of various frequencies of light is essential for applications ranging from window glass to fiber optic transmission cables and infrared (IR) heat-seeking missile detection systems. Light propagating through an optical system can be attenuated by absorption, reflection and scattering.

The interaction of light with matter can shed light on important information about the structure and dynamics of the material being examined. If the scattering centers are in motion, then the scattered radiation is Doppler shifted. An analysis of the spectrum of scattered light can thus yield information regarding the motion of the scattering center. Periodicity or structural repetition in the scattering medium will cause interference in the spectrum of scattered light. Thus, a study of the scattered light intensity as a function of scattering angle gives information about the structure, spatial configuration, or morphology of the scattering medium. With regards to light scattering in liquids and solids, primary material considerations include:[

- Crystalline structure: How close-packed its atoms or molecules are, and whether or not the atoms or molecules exhibit the *long-range order* evidenced in crystalline solids.
- Glassy structure: Scattering centers include fluctuations in density and/or composition.
- Microstructure: Scattering centers include internal surfaces in liquids due largely to density fluctuations, and microstructural defects in solids such as grains, grain boundaries, and microscopic pores.

In the process of light scattering, the most critical factor is the length scale of any or all of these structural features relative to the wavelength of the light being scattered.

An extensive review of light scattering in fluids has covered most of the mechanisms which contribute to the spectrum of scattered light in liquids, including density, anisotropy, and concentration fluctuations.[8] Thus, the study of light scattering by thermally driven density fluctuations (or Brillouin scattering) has been utilized successfully for the measurement of structural relaxation and viscoelasticity in liquids, as well as phase separation, vitrification and compressibility in glasses. In addition, the introduction of dynamic light scattering and photon correlation spectroscopy has made possible the measurement of the time dependence of spatial correlations in liquids and glasses in the relaxation time gap between 10^{-6} and 10^{-2} s in addition to even shorter time scales – or faster relaxation events. It has therefore become quite clear that light scattering is an extremely useful tool for monitoring the dynamics of structural relaxation in glasses on various temporal and spatial scales and therefore provides an ideal tool for quantifying the capacity of various glass compositions for guided light wave transmission well into the far infrared portions of the electromagnetic spectrum.

- Note: Light scattering in an ideal defect-free crystalline (non-metallic) solid which provides *no scattering centers* for incoming lightwaves will be due primarily to any effects of anharmonicity within the ordered lattice. Lightwave transmission will be highly directional due to the typical anisotropy of crystalline substances, which includes their symmetry group and Bravais lattice. For example, the seven different crystalline forms of quartz silica (silicon dioxide, SiO_2) are all clear, transparent materials.

Static and dynamic scattering

A common dichotomy in light scattering terminology is static light scattering versus dynamic light scattering. The chief distinction is whether the scattering is observed to be changing over time (dynamic) or constant over the observation (static). This terminology is especially commonly encountered in the field of polymer chemistry, though it can obviously be applied to a broad range of situations.

Dynamic light scattering (also known as **photon correlation spectroscopy** or **quasi-elastic light scattering**) is a technique in physics, which can be used to determine the size distribution profile of small particles in suspension or polymers in solution [1]. It can also be used to probe the behavior of complex fluids such as concentrated polymer solutions.

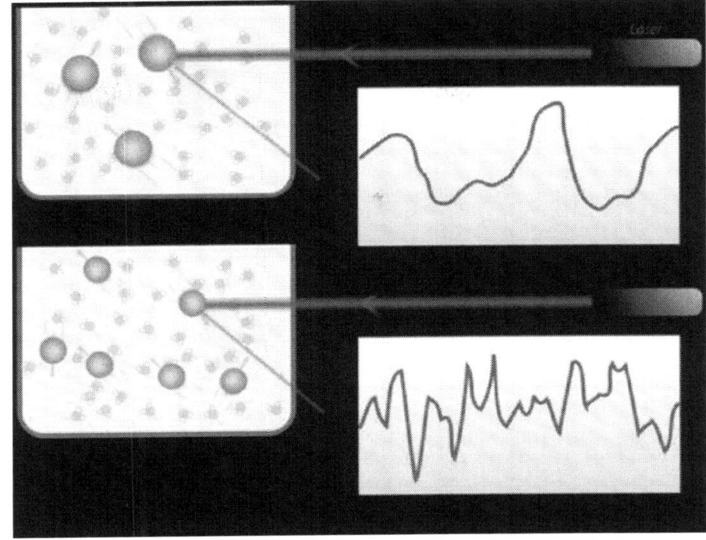

Fig. 17.5 : Dynamic light scattering

Static Light Scattering

Static Light scattering is a technique that uses intensity traces at number of angles to derive information about radius of gyration, molecular mass of the polymer complex and the second virial coefficient. There are typically a number of analyses developed to ana-lyze the scattering of particles in solution to derive the above physical characteristics of particles. The course of the scattered intensity as a function of the detector angle depends on size and structure of the particles.

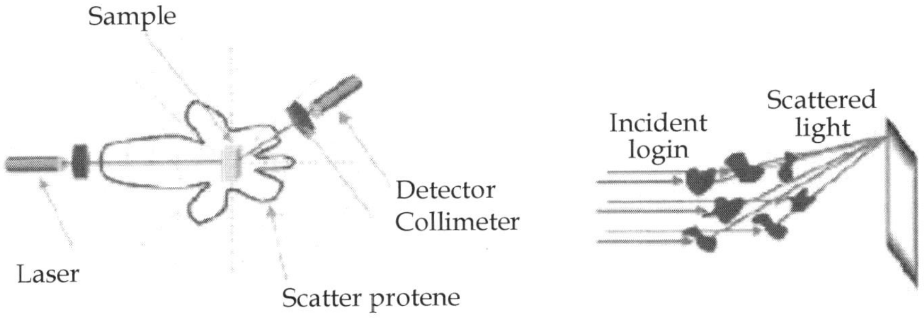

Fig. 17.6 : Basic principle and set up of static light scattering.

Rayleigh developed a theory that predicts the angular intensity distribution of light scattered by particles much smaller that the wavelength of light (R< »/20). Here the intensity of the scattered light depends only on the size of the particles and not on their concentration. If the intensity of the scattered light is measured in a plane orthogonal to the polarization plane of the incident light, there will be no angular distribution of the scattered light.

Dynamic Light Scattering

Another method to determine the size distribution of particles in solutions is the dynamic light scattering method. With this method we measure the variation of light intensity with time. In this method we usually pick an angle either 90° or 175° to conduct the measurements. The variation in time is caused by diffusion of particles which slightly changes scattering angle.

Relationship between diffusion constant and particle size (Stokes-Einstein law)

$$D \quad \frac{kT}{6\pi\eta r}$$

where k is the Boltzmann's constant, T is the temperature, r is the radius of the sphere and · is the viscosity of the fluid

Zeta Potential

Zeta potential is a measure of the magnitude of the repulsion or attraction between particles. Its measurement brings detailed insight into the dispersion mechanism and is the key to electrostatic dispersion control. In colloidal systems, zeta potential is used to represent the electrokinetic potential. Zeta potential can give a better understanding and control over colloidal suspensions.

Colloidal particles dispersed in a solution are electrically charged due to their ionic characteristics and dipolar attributes. Each particle dispersed in a solution is surrounded by oppositely charged ions called the fixed layer. Outside the fixed layer, there are vary-ing compositions of ions of opposite polarities, forming a cloud-like area. This area is called the diffuse double layer, and the whole area is electrically neutral.

When a voltage is applied to the solution in which particles are dispersed, particles are attracted to the electrode of the opposite polarity, accompanied by the fixed layer and part of the diffuse double layer, or internal side of the "sliding surface".

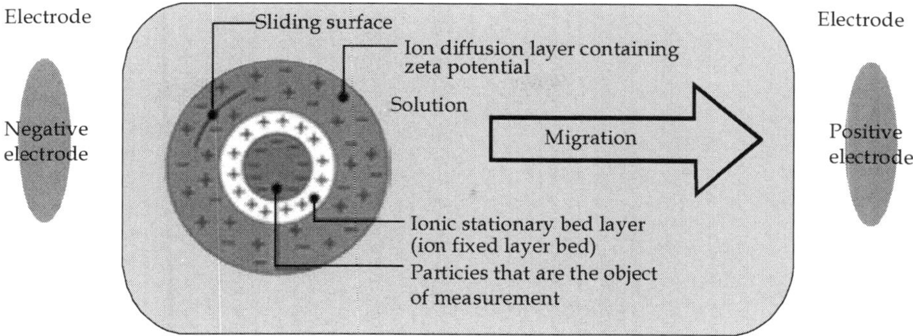

Fig. 17.7 : Formation of double layer in colloidal particles

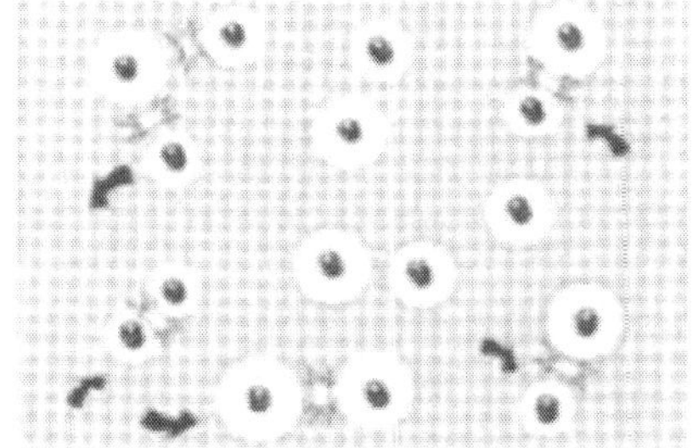

Fig. 17.8 : Charged particles repel each other

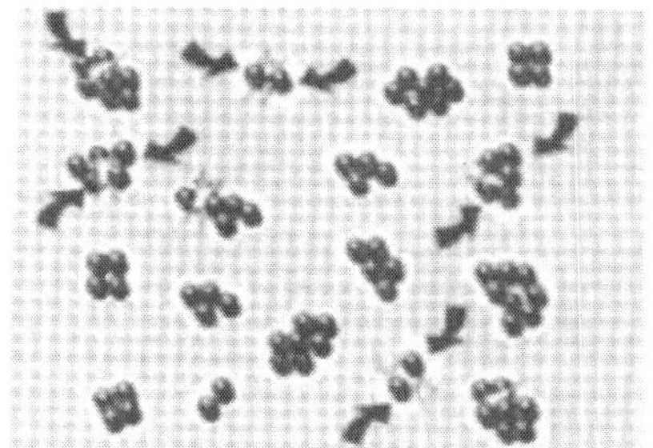

Fig. 17. 9 : Uncharged particles are free to collide and aggregate

The Zeta potential can be obtained by the Smoluchowski's formula, shown in equation. Smoluchowski theory is very powerful because it is valid for dispersed particles of any shape and any concentration.

$$\zeta = \frac{4\pi\eta}{\varepsilon} \times U \times 300 \times 300 \times 1000$$

ζ = Zeta Potential (mV)
η = Viscosity of Solution
ε = Dielectric Constant
$U = \frac{v}{V/L}$: Electrophoretic Mobility
v = Speed of Particle (cm / sec)
V = Voltage (V)
L = The distance of Electorode

❑❑❑

Chapter-18

Transmission Electron Microscopy

Electron Microscope

An electron microscope is a type of microscope that uses a particle beam of electrons to illuminate a specimen and create a highly-magnified image. Electron microscopes have much greater resolving power than light microscopes that use electromagnetic radiation and can obtain much higher magnifications of up to 1 million times, while the best light microscopes are limited to magnifications of 1000 times. Both electron and light microscopes have resolution limitations, imposed by the wavelength of the radiation they use. The greater resolution and magnification of the electron microscope is because the de Broglie wavelength of an electron is much smaller than that of a photon of visible light.

The electron microscope uses electrostatic and electromagnetic lenses in forming the image by controlling the electron beam to focus it at a specific plane relative to the specimen. This manner is similar to how a light microscope uses glass lenses to focus light on or through a specimen to form an image.

History

The first electron microscope prototype was built in 1931 by the German engineers Ernst Ruska and Max Knoll.[1] Although this initial instrument was capable of magnifying objects by only four hundred times, it demonstrated the principles of an electron microscope. Two years later, Ruska constructed an electron microscope that exceeded the resolution possible with an optical microscope.

Fig. 18.1 : Electron microscope constructed by Ernst Ruska in 1933

Reinhold Rudenberg, the scientific director of Siemens, had patented the electron microscope in 1931, stimulated by family illness to make the poliomyelitis virus particle visible. In 1937 Siemens began funding Ruska and Bodo von Borries to develop an electron microscope. Siemens also employed Ruska's brother Helmut to work on applications, particularly with biological specimens.

In the same decade Manfred von Ardenne pioneered the scanning electron microscope and his universal electron microscope.

Siemens produced the first commercial Transmission Electron Microscope (TEM) in 1939, but the first practical electron microscope had been built at the University of Toronto in 1938, by Eli Franklin Burton and students Cecil Hall, James Hillier, and Albert Prebus.

Although modern electron microscopes can magnify objects up to two million times, they are still based upon Ruska's prototype. The electron microscope is an essential item of equipment in many laboratories. Researchers use them to examine biological materials (such as microorganisms and cells), a variety of large molecules, medical biopsy samples, metals and crystalline structures and the characteristics of various surfaces. The electron microscope is also used extensively for inspection, quality assurance and failure analysis applications in industry, including, in particular, semiconductor device fabrication.

Types

Transmission Electron Microscope (TEM)

Scanning Electron Microscope (SEM)

Reflection Electron Microscope (REM)

Scanning Transmission Electron Microscope (STEM)

Low-voltage electron microscope (LVEM)

Applications

Semiconductor and data storage

- Circuit edit
- Defect analysis
- Failure analysis

Biology and life sciences

- Diagnostic electron microscopy
- Cryobiology
- Protein localization
- Electron tomography
- Cellular tomography
- Cryo-electron microscopy
- Toxicology
- Biological production and viral load monitoring
- Particle analysis
- Pharmaceutical QC

- Structural biology
- 3D tissue imaging
- Virology
- Vitrification

Research

- Electron beam-induced deposition
- Materials qualification
- Materials and sample preparation
- Nanoprototyping
- Nanometrology
- Device testing and characterization

Industry

- High-resolution imaging
- 2D & 3D micro-characterization
- Macro sample to nanometer metrology
- Particle detection and characterization
- Direct beam-writing fabrication
- Dynamic materials experiments
- Sample preparation
- Forensics
- Mining (mineral liberation analysis)
- Chemical/Petrochemical

Transmission electron microscopy

Transmission electron microscopy (TEM) is a microscopy technique whereby a beam of electrons is transmitted through an ultra thin specimen, interacting with the specimen as it passes through. An image is formed from the interaction of the electrons transmitted through the specimen; the image is magnified and focused onto an imaging device, such as a fluorescent screen, on a layer of photographic film, or to be detected by a sensor such as a CCD camera.

TEMs are capable of imaging at a significantly higher resolution than light microscopes, owing to the small de Broglie wavelength of electrons. This enables the instrument's user to examine fine detail-even as small as a single column of atoms, which is tens of thousands times smaller than the smallest resolvable object in a light microscope. TEM forms a major analysis method in a range of scientific fields, in both physical and biological sciences. TEMs find application in cancer research, virology, materials science as well as pollution, nanotechnology, and semiconductor research.

At smaller magnifications TEM image contrast is due to absorption of electrons in the material, due to the thickness and composition of the material. At higher magnifications complex wave interactions modulate the intensity of the image, requiring expert analysis of observed images. Alternate modes of use allow for the TEM to observe modulations in chemical identity, crystal orientation, electronic structure and sample induced electron phase shift as well as the regular absorption based imaging.

The first TEM was built by Max Knoll and Ernst Ruska in 1931, with this group developing the first TEM with resolving power greater than that of light in 1933 and the first commercial TEM in 1939.

The original form of electron microscope, the transmission electron microscope (TEM) uses a high voltage electron beam to create an image. The electrons are emitted by an electron gun, commonly fitted with a tungsten filament cathode as the electron source. The electron beam is accelerated by an anode typically at +100 keV (40 to 400 keV) with respect to the cathode, focused by electrostatic and electromagnetic lenses, and transmitted through the specimen that is in part transparent to electrons and in part scatters them out of the beam. When it emerges from the specimen, the electron beam carries information about the structure of the specimen that is magnified by the objective lens system of the microscope. The spatial variation in this information (the "image") is viewed by projecting the magnified electron image onto a fluorescent viewing screen coated with a phosphor or scintillator material such as zinc sulfide. The image can be photographically recorded by exposing a photographic film or plate directly to the electron beam, or a high-resolution phosphor may be coupled by means of a lens optical system or a fibre optic light-guide to the sensor of a CCD (charge-coupled device) camera. The image detected by the CCD may be displayed on a monitor or computer.

Resolution of the TEM is limited primarily by spherical aberration, but a new generation of aberration correctors have been able to partially overcome spherical aberration to increase resolution. Hardware correction of spherical aberration for the High Resolution TEM (HRTEM) has allowed the production of images with resolution below 0.5 Ångström (50 picometres) at magnifications above 50 million times. The ability to determine the positions of atoms within materials has made the HRTEM an important tool for nano-technologies research and development.

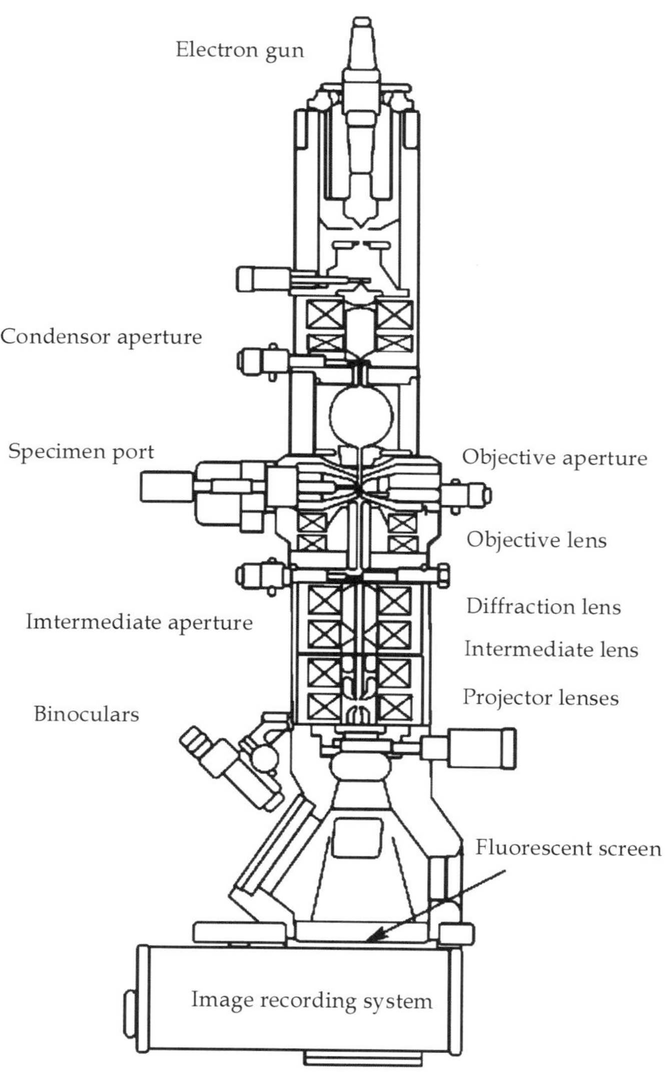

Fig. 18.2 : Transmission Electron Microscope

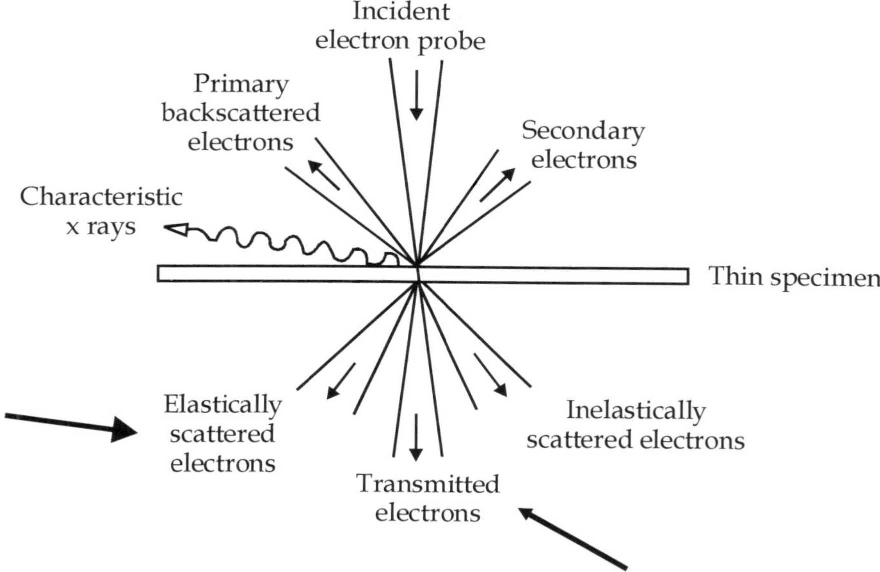

Fig. 18.3 : Interaction of electron beam and a thin specimen in TEM

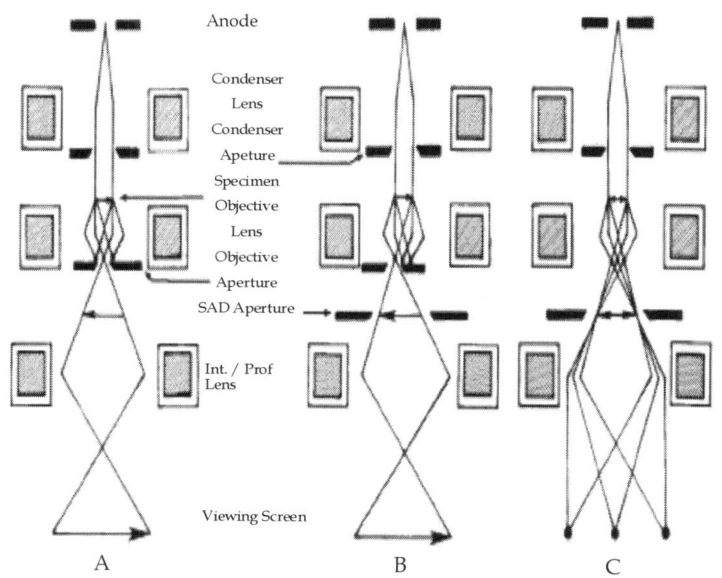

Configuration of a TEM for (A) BF imaging (B) DF imaging (C) observation of the electron diffraction pattern

Fig. 18.4 : Types of TEM image : Bright field (BF); Dark field (DF) & Diffraction Pattern

Mechanisms of image Formation

In TEM, the speciement is transparent for electrons; absorption of electrons plays a minor role in image formation. Deflection mechnism (scattering and diffraction of electrons) is mainly responsible for image formation.

Mass-thickness contrast (biology and polymer specimens)

Diffraction contrast (major image formation mechanisms)

Phase contrast (high resolution image of crystal lattice)

* Resolution Power : 0.2 nm at 1 MeV with 20 nm thick sample

Mass-thickness contrast

During passage of an electron wave through a specimen, scattering of electrons occurs at the atoms in this specimen. A certain portion of the incident electrons in deflected (scattered) in direction different from the promary beam direction.

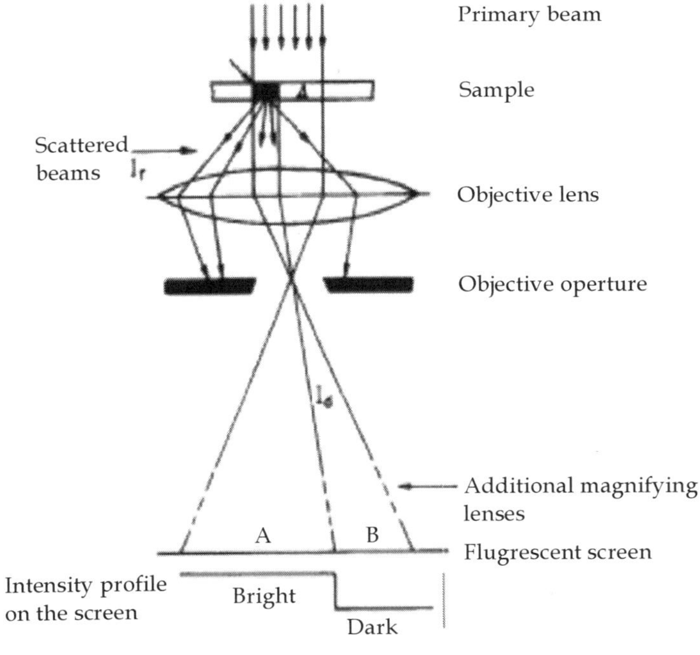

Fig. 18.5 : Scattering of electron beam

Mass-thickness contrast

The brightness of image is detemined by intensity of electron beam leaving the lower surface and pass through the objective aperture.

$I = I_0 \exp(-S^* p^*t)$

where, I is the transmitted beam intensity. I0 the incident beam intensity, p*t the mass thickness (p is the density and t is the thickness of the sample) and S the effective mass scattering cross-section, including size of objective aperture and incident beam energy, i.e. acceleration voltage.

The contrast, K, can be defined as

$K = \log(I0/I)$

The factors affecing electron scattering :

- Atomic number of specimen atoms
- Density of speciman
- Opening of objective apeture (size of aperture)
- Energy of electrons (acceleration voltage)

Using a objective aperture can enhance the mass-thickness contrast

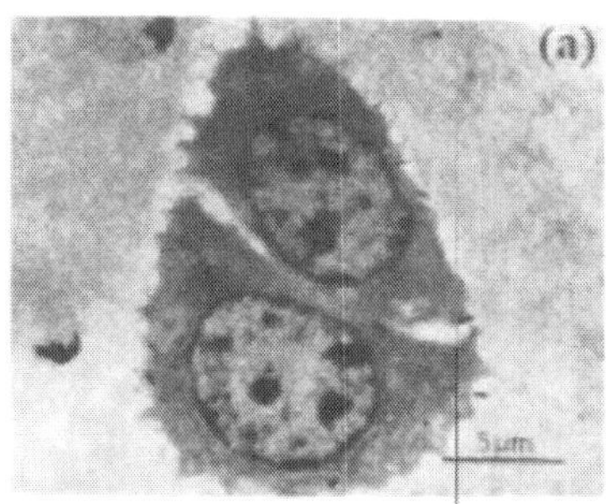

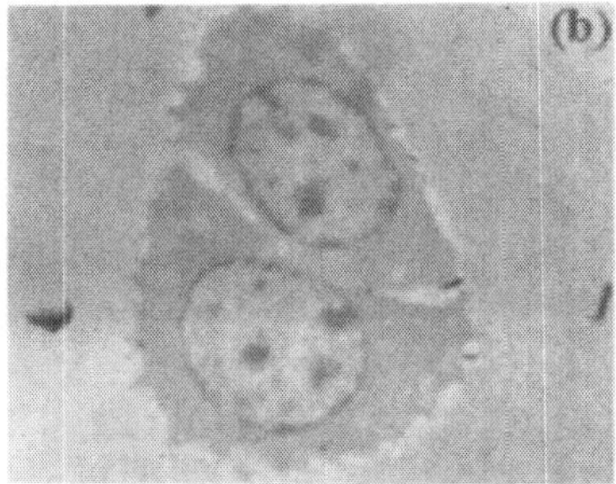

Fig. 18.6 : An animal cell photographed (a) with and (b) without an aperture in position

A* reduced objective aperture angel of scattering collection

A reduced beam voltage can enhance the mass-thickness contrast

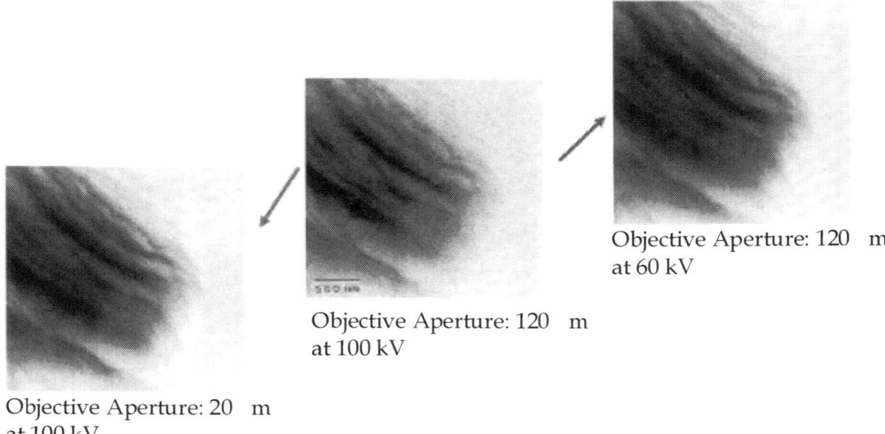

Fig. 18.7 : TEM micrographs of a PE thin film

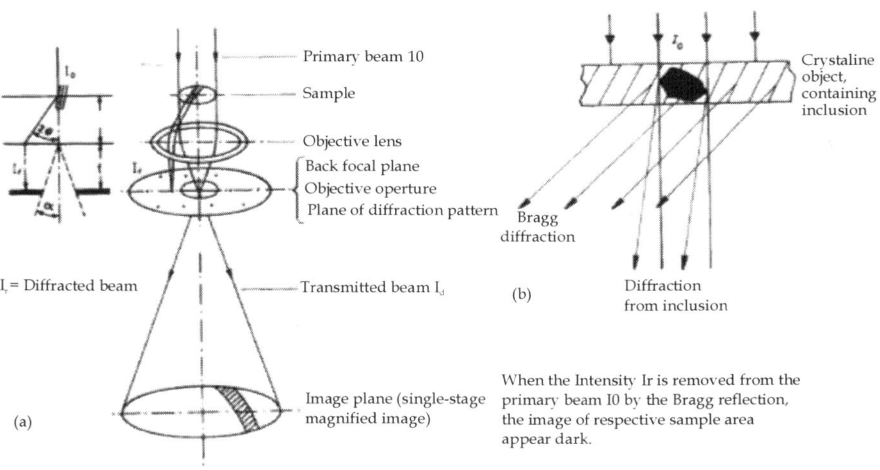

Fig. 18.8 : Diffraction contrast

Phase contrast

Both mass-thickness and diffraction contrast are based on the amplitude contrast mechanisms because they employ only the amplitudes of the scattered waves.

Phase contrast results from interference between waves of different phases.

- Most electron scattering involve a phase change
- Two or more diffracted beams are allowed to pass through the objective aperture.
- Each pair of beams, which interferes, will in principal give rise to a set of fringes in the images.

TEM micrograph with diffraction contrast of a semi-crystalline polymer thin film. The amorphous and crystalline phases are clearly seen.

Fig. 18.9 : Examples of diffraction contrast

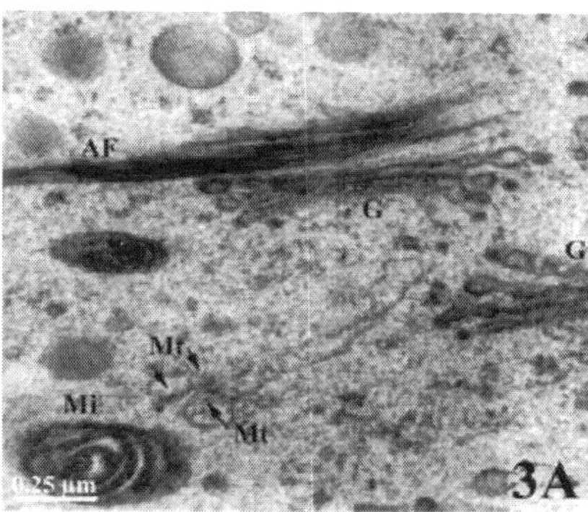

Fig. 18.10 : Organelles in a pollen grain of tobacco
(Nicotiana tabacum)
AF = Actin filaments;
G = Golgi apparatus;
Mi = Mitochondrion;
Mt =Microtubule

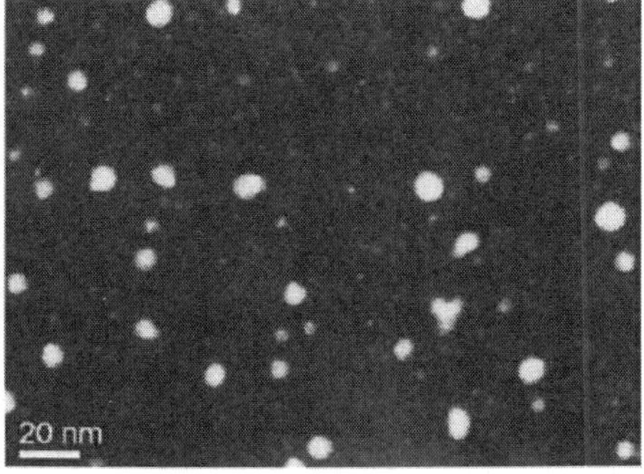

Fig. 18.11 : Gold nanoparticles inside to plant cell

TEM is mainly applied to study the inside of -sections of- very small object, e.g. membranes, nuclei and organelles of cells. But also extremely thin structures as a whole, like virussen and macromolecular aggregates (e.g. rings of porphyrin polymers), can be investigated with this instrument. The blackening in the final image is determined by the electron opacity of the object, which depends on its local thickness and the atomic composition of the material. A TEM provides thus different information than a scanning electron microscope (SEM) that produces images that tell something about the topography (surface) of the sample.

❑❑❑

Chapter-19

Scanning Electron Microscope

A Scanning electron microscope (SEM) is a type of electron microscope that images a sample by scanning it with a high-energy beam of electrons in a raster scan pattern. The electrons interact with the atoms that make up the sample producing signals that contain information about the sample's surface topography, composition, and other properties such as electrical conductivity.

The types of signals produced by an SEM include secondary electrons, back-scattered electrons (BSE), characteristic X-rays, light (cathodoluminescence), specimen current and transmitted electrons. Secondary electron detectors are common in all SEMs, but it is rare that a single machine would have detectors for all possible signals. The signals result from interactions of the electron beam with atoms at or near the surface of the sample. In the most common or standard detection mode, secondary electron imaging or SEI, the SEM can produce very high-resolution images of a sample surface, revealing details less than 1 nm in size. Due to the very narrow electron beam, SEM micrographs have a large depth of field yielding a characteristic three dimensional appearance useful for understanding the surface structure of a sample.

A wide range of magnification is possible, from about 10 times (about equivalent to that of a powerful hand-lens) to more than 500,000 times, about 250 times the magnification limit of the best light microscopes. Back-scattered electron (BSE) are beam electrons that are reflected from the sample by elastic scattering. BSE are often used in analytical SEM along with the spectra made from the characteristic X-rays. Because the intensity of the BSE signal is strongly related to the atomic number (Z) of the specimen, BSE images can provide information about the distribution of different elements in the sample. For the same reason, BSE imaging can image colloidal gold immune-labels of 5 or 10 nm diameter which would otherwise be difficult or impossible to detect in secondary electron images in biological specimens. Characteristic X-rays are emitted when the electron beam removes an inner shell electron from the sample, causing a higher energy electron to fill the shell and release energy. These characteristic X-rays are used to identify the composition and measure the abundance of elements in the sample.

In a typical SEM, an electron beam is thermionically emitted from an electron gun fitted with a tungsten filament cathode. Tungsten is normally used in thermionic electron guns because it has the highest melting point and lowest vapour pressure of all metals, thereby allowing it to be heated for electron emission, and because of its low cost. Other types of electron emitters include lanthanum hexaboride. (LAB_6) cathodes, which can be used in a standard tungsten filament SEM if the vacuum system is upgraded and field emission guns (FEG), which may be of the cold-cathode type using tungsten single crystal emitters or the thermally-assisted Schottky type, using emitters of zirconium oxide.

The electron beam, which typically has an energy ranging from 0.5 keV to 40 keV, is focused by one or two condenser lenses to a spot about 0.4 nm to 5 nm in diameter. The beam passes through pairs of scanning coils or pairs of deflector plates in the electron column, typically in the final lens, which deflect the beam in the x and y axes so that it scans in a raster fashion over a rectangular area of the sample surface.

When the primary electron beam interacts with the sample, the electrons lose energy by repeated random scattering and absorption within a teardrop-shaped volume of the specimen known as the interaction volume, which extends from less than 100 nm to around 5 μm into the surface. The size of the interaction volume depends on the electron's landing energy, the atomic number of the specimen and specimen's density. The energy exchange between the electron beam and the sample results in the reflection of high- energy electrons by elastic scattering, emission of secondary electrons by inelastic scattering and the emission of electromagnetic radiation, each of which can be detected by specialized detectors.

The beam current absorbed by the specimen can also be detected and used to create images of the distribution of specimen current. Electronic amplifiers of various types are used to amplify the signals which are displayed as variations in brightness on a cathode ray tube. The raster scanning of the CRT display is synchronised with that of the beam on the specimen in the microscope, and the resulting image is therefore a distribution map of the intensity of the signal being emitted from the scanned area of the specimen. The image may be captured by photography form a high resolution cathode ray tube, but in modern machines is digitally captured and displayed on a computer monitor and saved to a computer's hard disk.

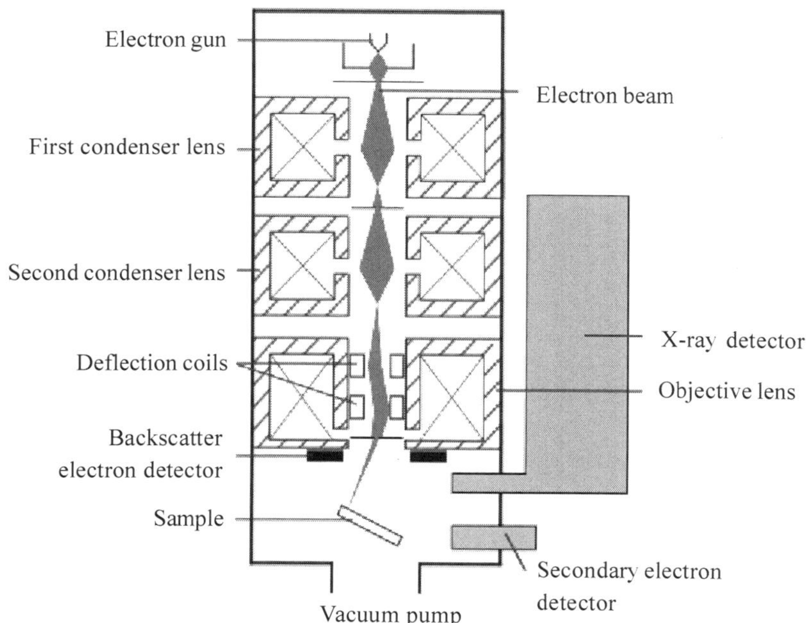

Fig. 19.1: Schematic Diagram of a SEM

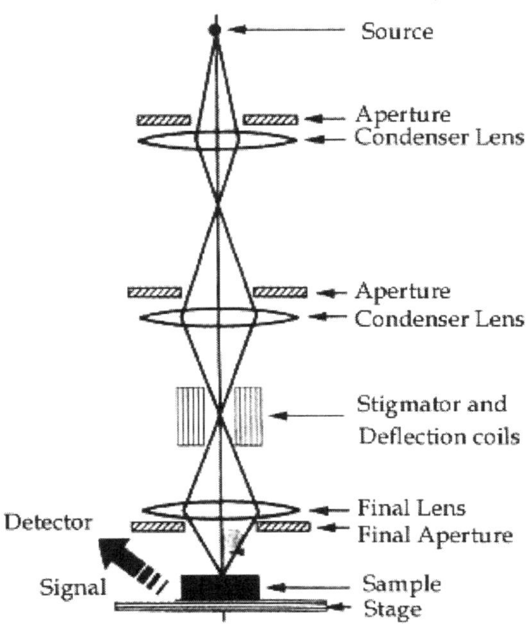

Fig. 19.2: Schematic diagram of a SEM

Magnification

Magnification in a SEM can be controlled over a range of up to 6 orders of magnitude from about 10 to 500,000 times. Unlike optical and transmission electron microscopes, image magnification in the SEM is not a function of the power of the objective lens. SEMs may have condenser and objective lenses, but their function is to focus the beam to a spot, and not to image the specimen. Provided the electron gun can generate a beam with sufficiently small diameter, a SEM could in principle work entirely without condenser or objective lenses, although it might not be very versatile or achieve very high resolution. In a SEM, as in scanning probe microscopy, magnification results from the ratio of the dimensions of the raster on the specimen and the raster on the display device. Assuming that the display screen has a fixed size, higher magnification results from reducing the size of the raster on the specimen, and vice versa. Magnification is therefore controlled by the current supplied to the x, y scanning coils, or the voltage supplied to the x, y deflector plates, and not by objective lens power.

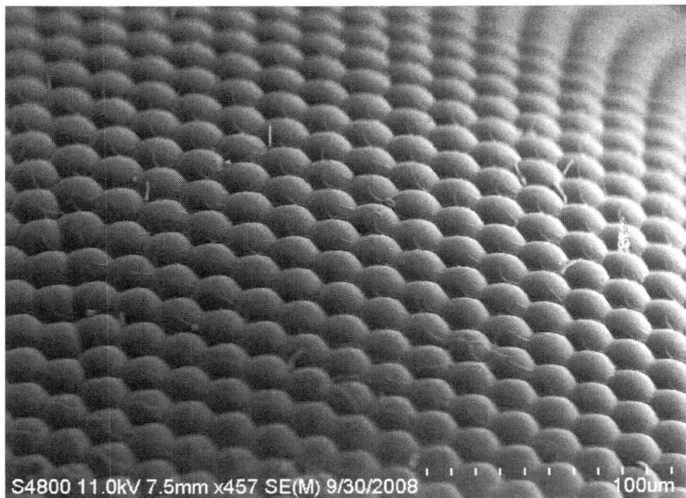

Fig. 19.3: An SEM micrograph of a house fly compound eye surface of 450 X magnification

Biological samples

For SEM, a specimen is normally required to be completely dry, since the specimen chamber is at high vacuum. Hard, dry materials such as wood, bone, feathers, dried insects or shells can be examined with little further treatment, but living cells and tissues and whole, soft bodied organisms usually require chemical fixation to preserve and stabilize their structure. Fixation is usually performed by incubation in a solution of a buffered chemical fixative, such as glutaraldehyde, sometimes in combination with formaldehyde and other fixatives and optionally followed by postfixation with osmium tetroxide. The fixed tissue is then dehydrated. Because air-drying causes collapse and shrinkage, this is commonly achieved by critical point drying, which involves replacement of water in the cells with organic solvents such as ethanol or acetone, and replacement of these solvents in turn with a transitional fluid such as liquid carbon dioxide at high pressure. The carbon dioxide is finally removed while in a supercritical state, so that no gas-liquid interface is present within the sample during drying. The dry specimen is usually mounted on a specimen stub using an adhesive, such as epoxy resin or electrically-conductive double-sided adhesive tape, and sputter coated with gold or gold/palladium alloy before examination in the microscope.

If the SEM is equipped with a cold stage for cryo-microscopy, cryofixation may be used and low-temperature scanning electrons microscopy performed on the cryogenically fixed specimens. Cryo-fixed specimens may be cryo-fractured under vacuum in a special apparatus to reveal internal structure, sputter coated and transferred onto the SEM cryo-stage while still frozen. Low-temperature scanning electron microscopy is also applicable to the imaging of temperature-sensitive materials such as iceand fats.

Freeze-fracturing, freeze-etch or freeze-and break is a preparation method particularly useful for examining lipid membranes and their incorporated proteins in "face on" view. The preparation method reveals the proteins embedded in the lipid bilayer.

Gold has a high atomic number and sputter coating with gold produced high topographic contrast and resolution. However, the coating has a thickness of a few nanometers, and can obscure the underlying fine detail of the specimen at very high magnification. Low, vacuum SEMs with differential pumping apertures allow samples to be imaged without such coating and without the loss of natural contrast caused by the coating, but are unable to achieve the resolution attainable by conventional SEMs with coated specimens.

Resolution of the SEM

The spatial resolution of the SEM depends on the size of the electron spot, which in turn depends on the both the wavelength of the electrons and the electron-optical system which produces the scanning beam. The resolution is also limited by the size of the interaction volume, or the extent to which the material interacts with the electron beam. The spot size and the interaction volume are both large compared to the distances between atoms, so the resolution of the SEM is not high enough to image individual atoms, as is possible in the shorter wavelength (i.e. higher energy) transmission electron microscope (TEM). The SEM has compensating advantages, though, including the ability to image a comparatively large area of the specimen; the ability to image bulk materials (not just thin films or foils); and the variety of analytical modes available for measuring the composition and properties of the specimen. Depending on the instrument, the resolution can fall somewhere between less than 1 nm and 20 nm. By 2009, The world's highest SEM resolution at high beam energies (0.4 nm at 30 kV) is obtained with the Hitachi S-5500. At low beam energies, the best resolution (by 2009) is achieved by the Magellan XHR system for FEI Company (0.9 nm to 1 kV).

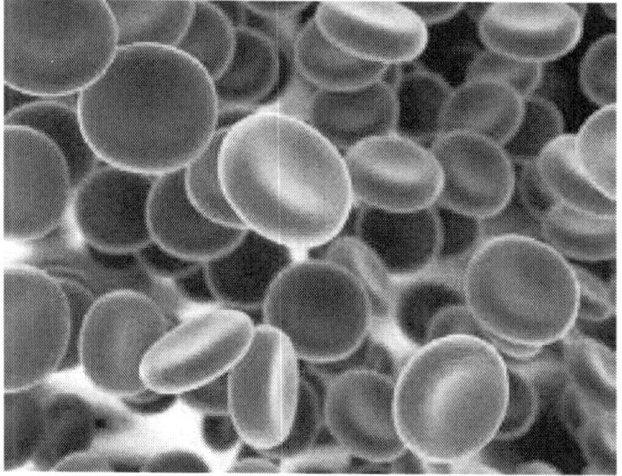

Fig. 19.4: Red-blood-cells

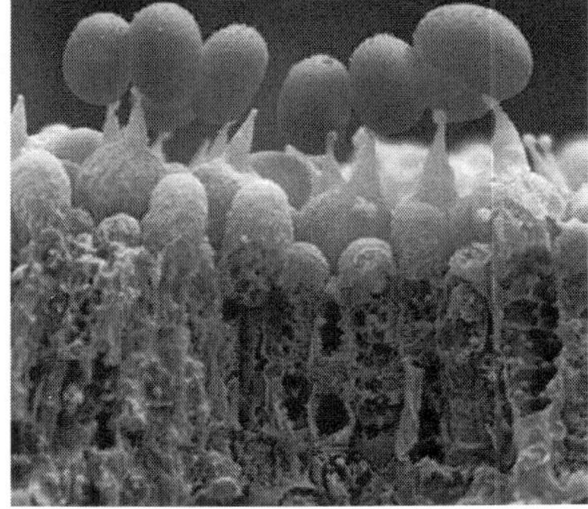

Fig. 19.5: Mushrooms spores

Chapter-20

Infra-Red and NMR Spectroscopy

Infrared spectroscopy (IR spectroscopy) is the spectroscopy that deals with the infrared region of the electromagnetic spectrum, that is light with a longer wavelength and lower frequency than visible light. It covers a range of techniques, mostly based on absorption spectroscopy. As with all spectroscopic techniques, it can be used to identify and study chemicals. A common laboratory instrument that uses this technique is a Fourier transform infrared (FTIR) spectrometer.

The infrared portion of the electromagnetic spectrum is usually divided into three regions; the near-, mid- and far- infrared, named for their relation to the visible spectrum. The higher energy near-IR, approximately 14000–4000 cm^{-1} (0.8–2.5 μm wavelength) can excite overtone or harmonic vibrations. The mid-infrared, approximately 4000–400 cm^{-1} (2.5–25 μm) may be used to study the fundamental vibrations and associated rotational-vibrational structure. The far-infrared, approximately 400–10 cm^{-1} (25–1000 μm), lying adjacent to the microwave region, has low energy and may be used for rotational spectroscopy. The names and classifications of these sub regions are conventions, and are only loosely based on the relative molecular or electromagnetic properties.

Infrared spectroscopy exploits the fact that molecules absorb specific frequencies that are characteristic of their structure. These absorptions are resonant frequencies, i.e. the frequency of the absorbed radiation matches the frequency of the bond or group that vibrates. The energies are determined by the shape of the molecular potential energy surfaces, the masses of the atoms, and the associated vibronic coupling.

The infrared spectrum of a sample is recorded by passing a beam of infrared light through the sample. When the frequency of the IR is the same as the vibrational frequency of a bond, absorption occurs. Examination of the transmitted light reveals how much energy was absorbed at each wavelength. This can be done with a monochromatic beam, which changes in wavelength over time, or by using a Fourier transform instrument to measure all wavelengths at once. From this, a transmittance or absorbance spectrum can be produced, showing at which IR wavelengths the sample absorbs. Analysis of these absorption characteristics reveals details about the molecular structure of the sample.

This technique works almost exclusively on samples with covalent bonds. Simple spectra are obtained from samples with few IR active bonds and high levels of purity. More complex molecular structures lead to more absorption bands and more complex spectra. The technique has been used for the characterization of very complex mixtures.

A technique in which infrared light is passed through matter and some of the light is absorbed by inciting molecular vibration. The difference between the incident and the emitted radiation reveals structural and functional data about the molecule.

Infrared or thermal imaging is a heat-sensing technology that can see in complete darkness without the need for illumination. All objects emit infrared energy, but it cannot be seen by the human eye without the use of technology. Infrared is a part of the electromagnetic spectrum (see diagram below) with wavelengths that are longer than visible light.

The warmer an object, the more energy it emits. Infrared energy emitted by a viewed scene is focused through the specialized objective lens assembly of an infrared camera on to the camera's focal plane array (FPA). The FPA uses materials that respond by generating electrical impulses when infrared energy strikes it. These electrical impulses are then sent in the form of temperature values to an image signal processor that turns them into video data for presentation on a display.

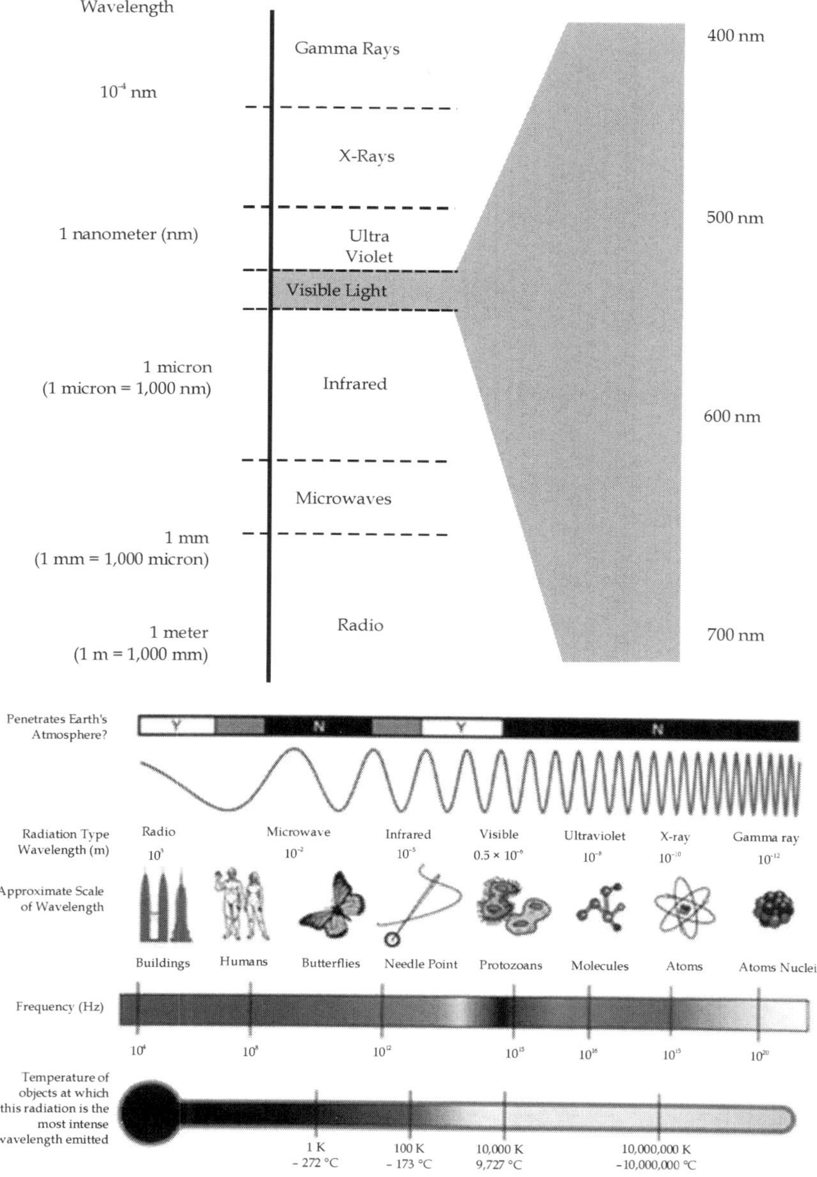

Fig. 20.1 : EM Spectrum chart

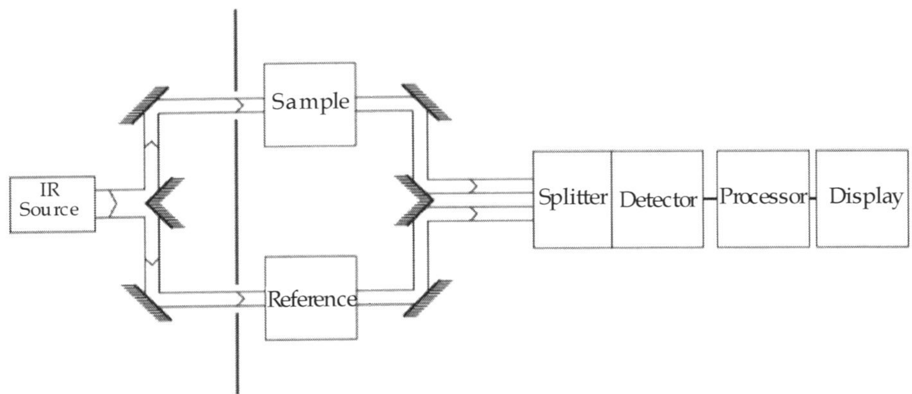

Fig. 20.2 : Schematics of a two-beam absorption spectrometer.

Fourier transform infrared (FTIR) spectroscopy

Fourier transform infrared (FTIR) spectroscopy is a measurement technique that allows one to record infrared spectra. Infrared light is guided through an interferometer and then through the sample (or vice versa). A moving mirror inside the apparatus alters the distribution of infrared light that passes through the interferometer. The signal directly recorded, called an "interferogram", represents light output as a function of mirror position. A data-processing technique called Fourier transform turns this raw data into the desired result (the sample's spectrum): Light output as a function of infrared wavelength (or equivalently, wavenumber). As described above, the sample's spectrum is always compared to a reference.

There is an alternate method for taking spectra (the "dispersive" or "scanning monochromator" method), where one wavelength at a time passes through the sample. The dispersive method is more common in UV-Vis spectroscopy, but is less practical in the infrared than the FTIR method. One reason that FTIR is favored is called "Fellgett's advantage" or the "multiplex advantage": The information at all frequencies is collected simultaneously, improving both speed and signal-to-noise ratio. Another is called "Jacquinot's Throughput Advantage": A dispersive measurement requires detecting much lower light levels than an FTIR measurement. There are other advantages, as well as some disadvantages,[2] but virtually all modern infrared spectrometers are FTIR instruments.

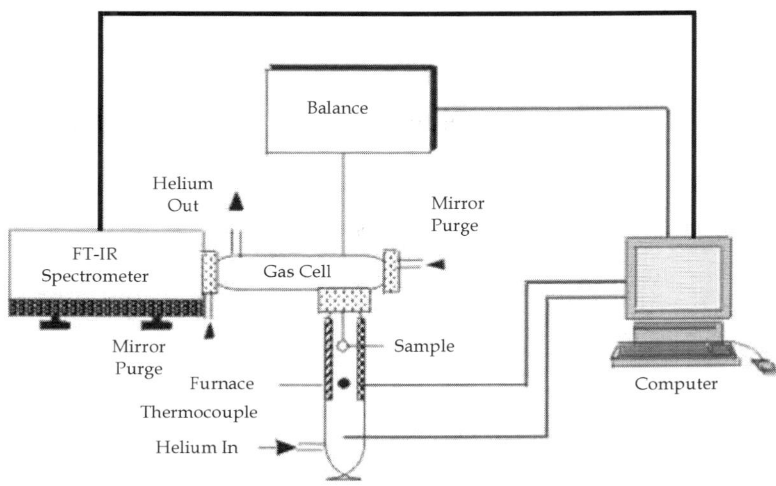

Fig. 20.3 : FTIR Spectroscopy

Uses and applications

Infrared spectroscopy is a simple and reliable technique widely used in both organic and inorganic chemistry, in research and industry. It is used in quality control, dynamic measurement, and monitoring applications such as the long-term unattended measurement of CO_2 concentrations in greenhouses and growth chambers by infrared gas analyzers.It is also used in forensic analysis in both criminal and civil cases, for example in identifying polymer degradation. It can be used in detecting how much alcohol is in the blood of a suspected drink driver measured as 1/10,000 g/mL = 100 μg/mL.

A useful way of analysing solid samples without the need for cutting samples uses ATR or attenuated total reflection spectroscopy. Soft samples such as elastomers are pressed against a selenium single crystal and the spectrum measured by reflection form the outermost sample layers.

With increasing technology in computer filtering and manipulation of the results, samples in solution can now be measured accurately (water produces a broad absorbance across the range of interest, and thus renders the spectra unreadable without this computer treatment).

Some instruments will also automatically tell you what substance is being measured from a store of thousands of reference spectra held in storage.

Infrared spectroscopy is also useful in measuring the degree of polymerization in polymer manufacture. Changes in the character or quantity of a particular bond are assessed by measuring at a specific frequency over time. Modern research instruments can take infrared measurements across the range of interest as frequently as 32 times a second. This can be done whilst simultaneous measurements are made using other techniques. This makes the observations of chemical reactions and processes quicker and more accurate.

Infrared spectroscopy has also been successfully utilized in the field of semiconductor microelectronics: for example, infrared spectroscopy can be applied to semiconductors like silicon, gallium arsenide, gallium nitride, zinc selenide, amorphous silicon, silicon nitride, etc.

The instruments are now small, and can be transported, even for use in field trials.

Nanolevel

Infrared spectroscopy is a characterization technique that is widely used to observe molecular vibration signatures. Using a particular optical configuration such as multiple internal reflections (MIR) or attenuated total reflection (ATR) it is possible to enhance the signal from ultra thin layers or from low concentration contaminants. Reasons of the enhancement of the sensitivity, thanks to these optical configurations, are explained. Advantages of these techniques are highlighted by showing a wide range of applications in which they are used. We present here, the results obtained using these configurations in several fields of microelectronics or nanotechnology: i) the silanization of silicon surfaces are being developed for biological applications; a study of the grafting efficiency and verification of different processes which modify the terminal functional groups are shown; ii) the characterization of high materials designed to replace silicon gate oxides 32 nm technology node; results obtained from wet etching of HfO_2 are presented and the correlation between the wet etching ratio and the morphology of the layer is pointed out; iii) the study of porous SiOC materials used as new interconnection insulators; the mechanical stability of the stacks including low is critical and the identification of the fracture interfaces obtained using 4 points bending tests is investigated;. ©2007 *American Institute of Physics*

Nuclear magnetic resonance spectroscopy

Nuclear magnetic resonance (**NMR**) is an effect whereby magnetic nuclei in a magnetic field absorb and re-emit electromagnetic (EM) energy. This energy is at a specific resonance frequency which depends on the strength of the magnetic field and other factors. This allows the observation of specific quantum mechanical magnetic properties of an atomic nucleus. Many scientific techniques exploit NMR phenomena to study molecular physics, crystals and non-crystalline materials through NMR spectroscopy. NMR is also routinely used in advanced medical imaging techniques, such as in magnetic resonance imaging (MRI).

All stable isotopes that contain an odd number of protons and/or of neutrons (see Isotope) have an intrinsic magnetic moment and angular momentum, in other words a nonzero spin, while all nuclides with even numbers of both have spin 0. The most commonly studied nuclei are 1

H (the most NMR-sensitive isotope after the radioactive 3 H) and 13 C, although nuclei from isotopes of many other elements (e.g. ^{2}H, 10 B, 11 B, 14 N, 15 N, 17 O, 19 F, 23 Na, 29 Si, 31P, 35 Cl, 113 Cd, 129 Xe, 195 Pt) are studied by high-field NMR spectroscopy as well.

A key feature of NMR is that the resonance frequency of a particular substance is directly proportional to the strength of the applied magnetic field. It is this feature that is exploited in imaging techniques; if a sample is placed in a non-uniform magnetic field then the resonance frequencies of the sample's nuclei depend on where in the field they are located. Since the resolution of the imaging technique depends on the magnitude of magnetic field gradient, many efforts are made to develop increased field strength, often using superconductors. The effectiveness of NMR can also be improved using hyperpolarization, and/or using two-dimensional, three-dimensional and higher-dimensional multi-frequency techniques.

The principle of NMR usually involves two sequential steps:

- The alignment (polarization) of the magnetic nuclear spins in

  O·

- The perturbation of this alignment of the nuclear spins by employing an electro-magnetic, usually radio frequency (RF) pulse. The required perturbing frequency is dependent upon the static magnetic field (H_O) and the nuclei of observation.

The two fields are usually chosen to be perpendicular to each other as this maximizes the NMR signal strength. The resulting response by the total magnetization (M) of the nuclear spins is the phenomenon that is exploited in NMR spectroscopy and magnetic resonance imaging. Both use intense applied magnetic fields (H_0) in order to achieve dispersion and very high stability to deliver spectral resolution, the details of which are described by chemical shifts, the Zeeman effect, and Knight shifts (in metals).

NMR phenomena are also utilized in low-field NMR, NMR spectroscopy and MRI in the Earth's magnetic field (referred to as Earth's field NMR), and in several types of magnetometers.

Theory of nuclear magnetic resonance

Nuclear spin and magnets

All nucleons, that is neutrons and protons, composing any atomic nucleus, have the intrinsic quantum property of spin. The overall spin of the nucleus is determined by the spin quantum number *S*. If the number of both the protons and neutrons in a given nuclide are even then $S = 0$, i.e. there is no overall spin; just as electrons pair up in atomic orbitals, so do even numbers of protons or even numbers of neutrons (which are also spin-$^1D_2$ particles and hence fermions) pair up giving zero overall spin.

However, a proton and neutron will have lower energy when their spins are parallel, not anti-parallel, as this parallel spin alignment does not infringe upon the Pauli principle, but instead has to do with the quark structure of these two nucleons. Therefore, the spin ground state for the deuteron (the deuterium nucleus, or the ^2H isotope of hydrogen)—that has only a proton and a neutron—corresponds to a spin value of 1, not of zero; the single, isolated deuteron is therefore exhibiting an NMR absorption spectrum characteristic of a quadrupolar nucleus of spin 1, which in the 'rigid' state at very low temperatures is a characteristic ('Pake') doublet, (not a singlet as for a single, isolated ^1H, or any other isolated fermion or dipolar nucleus of spin 1/2). On the other hand, because of the Pauli principle, the (radioactive) tritium isotope has to have a pair of anti-parallel spin neutrons (of total spin zero for the neutron spin couple), plus a proton of spin 1/2; therefore, the character of the tritium nucleus ('triton') is again magnetic dipolar, *not quadrupolar*—like its non-radioactive deuteron neighbor—and the tritium nucleus total spin value is again 1/2, just like for the simpler, abundant hydrogen isotope, ^1H nucleus (the *proton*).

The NMR absorption (radio) frequency for tritium is however slightly higher than that of ^1H because the tritium nucleus has a slightly higher gyromagnetic ratio than ^1H. In many other cases of *non-radioactive* nuclei, the overall spin is also non-zero. For example, the 27 Al nucleus has an overall spin value $S = 5/2$.

A non-zero spin is thus always associated with a non-zero magnetic moment (i) via the relation $i = aS$, where ^{3} is the gyromagnetic ratio. It is this magnetic moment that allows the observation of NMR absorption spectra caused by transitions between nuclear spin levels. Most nuclides (with some rare exceptions) that have both even numbers of protons and even numbers of neutrons, also have zero nuclear magnetic moments-and also have zero magnetic dipole and quadrupole moments; therefore, such nuclides do not exhibit any NMR absorption spectra. Thus, 18 O is an example of a nuclide that has no NMR absorption, whereas 13C, 31 P, 35 Cl and 37 Cl are nuclides that do exhibit NMR absorption spectra; the last two nuclei are quadrupolar nuclei whereas the preceding two nuclei (13 C and 31 P) are dipolar ones.

Electron spin resonance (ESR) is a related technique which detects transitions between electron spin levels instead of nuclear ones. The basic principles are similar but the instrumentation, data analysis and detailed theory are significantly different. Moreover, there is a much smaller number of molecules and materials with unpaired electron spins that exhibit ESR (or electron paramagnetic resonance (EPR)) absorption than those that have NMR absorption spectra. ESR has much higher sensitivity than NMR.

Values of spin angular momentum

The angular momentum associated with nuclear spin is quantized. This means both that the magnitude of angular momentum is quantized (i.e. S can only take on a restricted range of values), and also that the orientation of the associated angular momentum is quantized. The associated quantum number is known as the magnetic quantum number, m, and can take values from $+S$ to S, in integer steps. Hence for any given nucleus, there is a total of $2S + 1$ angular momentum states.

The z-component of the angular momentum vector (S) is therefore S_z = m′, where ′ is the reduced Planck constant. The z-component of the magnetic moment is simply:

Magnetic resonance by nuclei

Resonant absorption by nuclear spins will occur only when electromagnetic radiation of the correct frequency (e.g., equaling the Larmor precession rate) is being applied to match the energy difference between the nuclear spin levels in a constant magnetic field of the appropriate strength. The energy of an absorbed photon is then $E = h½_0$, where $½_0$ is the resonance radiofrequency that has to match (that is, it has to be equal to) the Larmor precession frequency $½_L$ of the nuclear magnetization in the constant magnetic field B_0. Hence, a magnetic resonance absorption will only occur when "E = $h½_0$, which is when $½_0$ = $^3B_0/(2À)$. Such magnetic resonance frequencies typically correspond to the radio frequency (or RF) range of the electromagnetic spectrum for magnetic fields up to ~20 T. It is this magnetic resonant absorption which is detected in NMR.

NMR spectroscopy

NMR spectroscopy is one of the principal techniques used to obtain physical, chemical, electronic and structural information about molecules due to either the chemical shift, Zeeman effect, or the Knight shift effect, or a combination of both, on the resonant frequencies of the nuclei present in the sample. It is a powerful technique that can provide detailed information on the topology, dynamics and three-dimensional structure of molecules in solution and the solid state. Thus, structural and dynamic information is obtainable (with or without "magic angle" spinning (MAS)) from NMR studies of quadrupolar nuclei (that is, those nuclei with spin S > $^1D_2$) even in the presence of magnetic dipole-dipole interaction broadening (or simply, dipolar broadening) which is always much smaller than the quadrupolar interaction strength because it is a magnetic vs. an electric interaction effect.

Additional structural and chemical information may be obtained by performing double-quantum NMR experiments for quadrupolar nuclei such as 2 H. Also, nuclear magnetic resonance is one of the techniques that has been used to design quantum automata, and also build elementary quantum computers.

NMR

- NMR is the most powerful tool available for organic structure determination.
- It is used to study a wide variety of nuclei: ^1H, ^{13}C, ^{15}N, ^{19}F, ^{31}P

Nuclear Spin

- A nucleus with an odd atomic number or an odd mass number has a nuclear spin.
- The spinning charged nucleus generates a magnetic field.

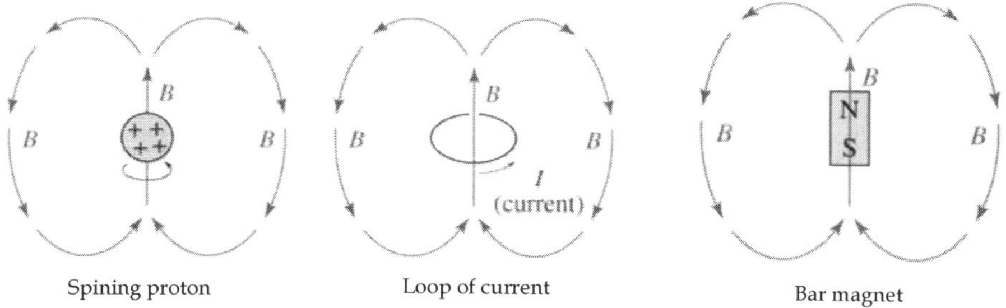

Fig. 20.4 : Images of nuclear spin

External Magnetic Field

When placed in an external field, spinning protons act like bar magnets.

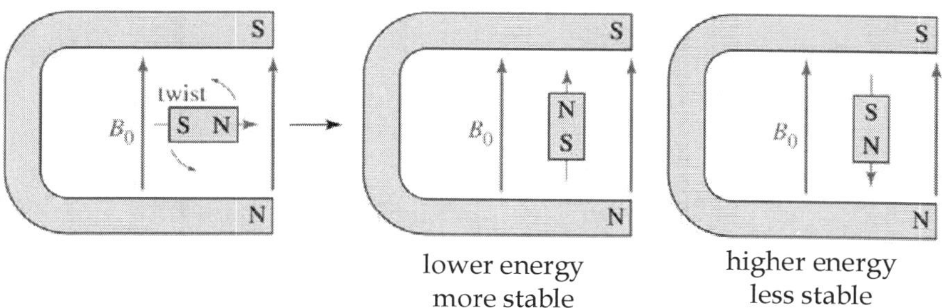

Fig. 20.5 : Proton magnetic field

Two Energy States

The magnetic fields of the spinning nuclei will align either *with* the external field, or *against* the field. A photon with the right amount of energy can be absorbed and cause the spinning proton to flip.

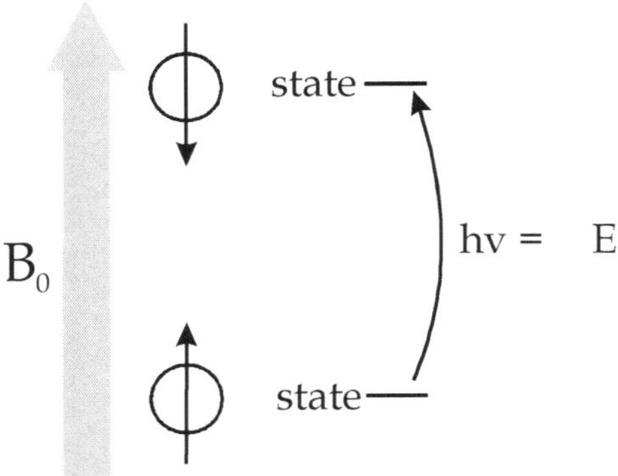

Fig. 20.6 : Energy states

Protons in a Molecule

Depending on their chemical environment, protons in a molecule are shielded by different amounts.

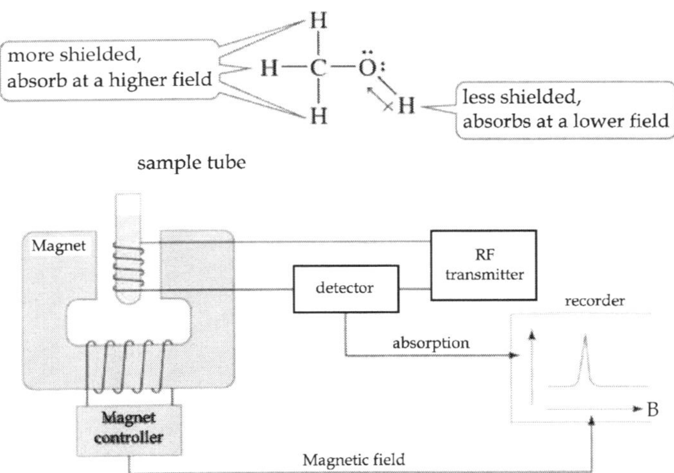

Fig. 20.7 : The NMR Spectrometer

NMR Signals

- The *number* of signals shows how many different kinds of protons are present.
- The *location* of the signals shows how shielded or deshielded the proton is.

- The *intensity* of the signal shows the number of protons of that type.
- Signal *splitting* shows the number of protons on adjacent atoms.

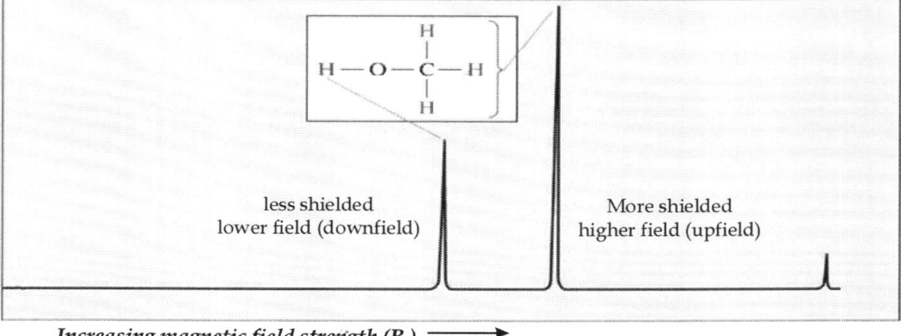

Fig. 20.8 : The NMR Graph

Chemical Shift

- Measured in parts per million.
- Ratio of shift downfield from TMS (Hz) to total spectrometer frequency (Hz).
- Same value for 60, 100, or 300 MHz machine.
- Called the delta scale.

$$\text{chemical shift, ppm } \delta = \frac{\text{shift downfield from TMS (in Hz)}}{\text{spectrometer frequency (in MHz)}}$$

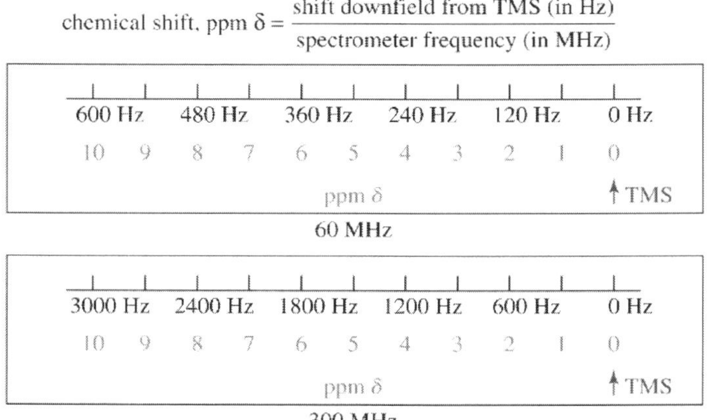

Fig. 20.9 : Delta Scale

Location of Signals

- More electronegative atoms deshield more and give larger shift values.
- Effect decreases with distance.
- Additional electronegative atoms cause increase in chemical shift.

O-H and N-H Signals

- Chemical shift depends on concentration.
- Hydrogen bonding in concentrated solutions deshield the protons, so signal is around d3.5 for N-H and d4.5 for O-H.
- Proton exchanges between the molecules broaden the peak.

Identifying the O-H or N-H Peak

- Chemical shift will depend on concentration and solvent.
- To verify that a particular peak is due to O-H or N-H, shake the sample with D_2O
- Deuterium will exchange with the O-H or N-H protons.
- On a second NMR spectrum the peak will be absent, or much less intense.

Carbon-13

- ^{12}C has no magnetic spin.
- ^{13}C has a magnetic spin, but is only 1% of the carbon in a sample.
- The gyromagnetic ratio of 13C is one-fourth of that of 1H.
- Signals are weak, getting lost in noise.
- Hundreds of spectra are taken, averaged.

NMR applications in Nanotechnology

Buckminsterfullerene C_{60} was first detected in the time-of-flight mass spectrometer from the product of laser vaporization of graphite1 and later obtained in large quantity from electric arc resistive heating of graphite rods.2 & 32, 3 The four characteristic peaks in the infrared (IR) spectrum of C_{60} revealed its truncated icosahedron structure,3 which was further confirmed by nuclear magnetic resonance (NMR) spectroscopy.4-64, 5, 6 Separation and characterization of fullerenes C_{70},4 & 54, 5, 7 C_{76},8 C_{78},9-129, 10, 11, 12 C_{82},11 and those with even higher carbon atom counts followed. High-performance liquid chromatography (HPLC) has proved to be a powerful technique to separate fullerene ^{13}C NMR is the method of choice to characterize them.

The cage-like structure of a fullerene consists of 12 five-membered rings and a number of six-membered rings depending on the number of carbon atoms. It was postulated that the five-membered rings need to be separated from each other to reduce the localization of the strain caused by the bending of the sp^2-hybridized carbon atoms. This is the essence of the isolated pentagon rule (IPR), which dramatically reduces the possible number of isomers for a given fullerene family. Fig. 20.10

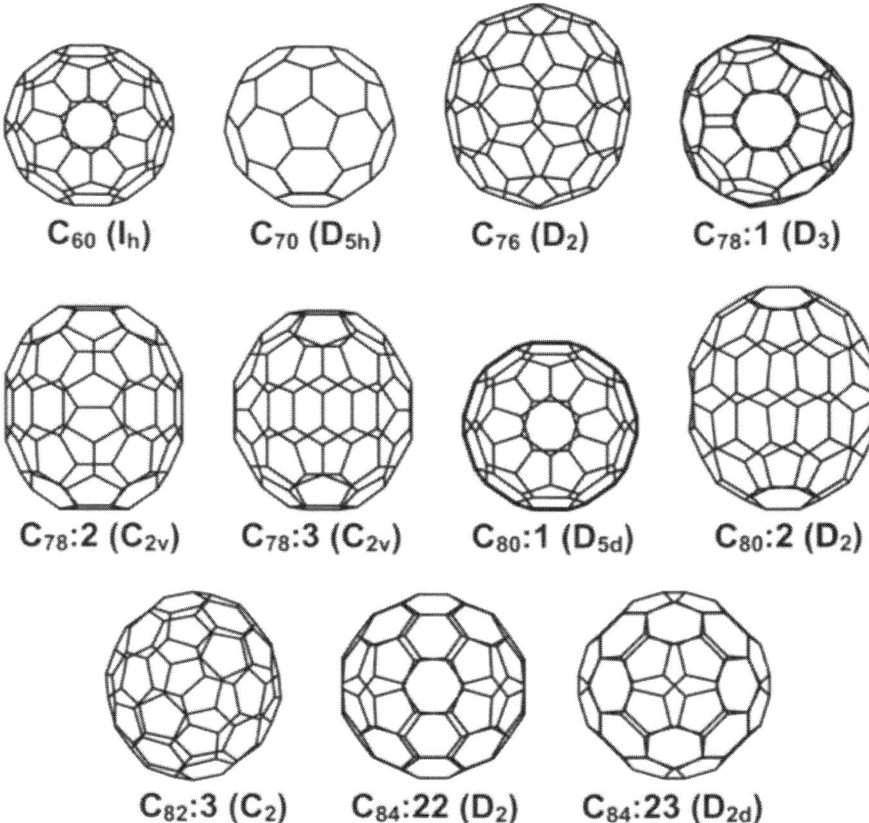

Fig. 20.10 : The major isomers of fullerenes C_{60}, C_{70}, C_{76}, C_{80}, C_{82}, and C_{84}.

shows the major isomers of fullerenes C_{60} to C_{84}. On the other hand, it is still common that several IPR isomers of a fullerene have same point-group symmetry and have same ^{13}C NMR patterns. For instance, isomers 1, 5, 21, and 22 of fullerene C_{84} all have D_2 symmetry and give 21 NMR peaks with equal intensity. One of the two major isomers of C_{84} has D_2 symmetry as observed in experiment, but its structure cannot be determined from the one-dimensional (1-D) NMR spectrum. Relative energies calculated using quantum mechanics are helpful, but not definitive in determination of the molecular structure. Accurate theoretical chemical shifts predicted by high-level quantum mechanical calculations have been shown to result in unambiguous assignment for those isomers.

❑❑❑

Chapter-21

Nanotechnology in Agriculture

Nanotechnology has the potential to revolutionize the agricultural and food industry with new tools for the molecular treatment of diseases, rapid disease detection, enhancing the ability of plants to absorb nutrients etc. Smart sensors and smart delivery systems will help the agricultural industry combat viruses and other crop pathogens. In the near future nanostructured catalysts will be available which will increase the efficiency of pesticides and herbicides, allowing lower doses to be used. Nanotechnology will also protect the environment indirectly through the use of alternative (renewable) energy supplies, and filters or catalysts to reduce pollution and clean-up existing pollutants.

Precision Farming

Precision farming has been a long-desired goal to maximise output (i.e. crop yields) while minimising input (i.e. fertilisers, pesticides, herbicides, etc) through monitoring environmental variables and applying targeted action. Precision farming makes use of computers, global satellite positioning systems, and remote sensing devices to measure highly localised environmental conditions thus determining whether crops are growing at maximum efficiency or precisely identifying the nature and location of problems. By using centralised data to determine soil conditions and plant development, seeding, fertilizer, chemical and water use can be fine-tuned to lower production costs and potentially increase production- all benefiting the farmer.Precision farming can also help to reduce agricultural waste and thus keep environmental pollution to a minimum. Although not fully implemented yet, tiny sensors and monitoring systems enabled by nanotechnology will have a large impact on future precision farming methodologies.

- Nanomolecules adsorbed on clay particles
- Prevent fixation of nutrient ions
- Favours labile pool of nutrients

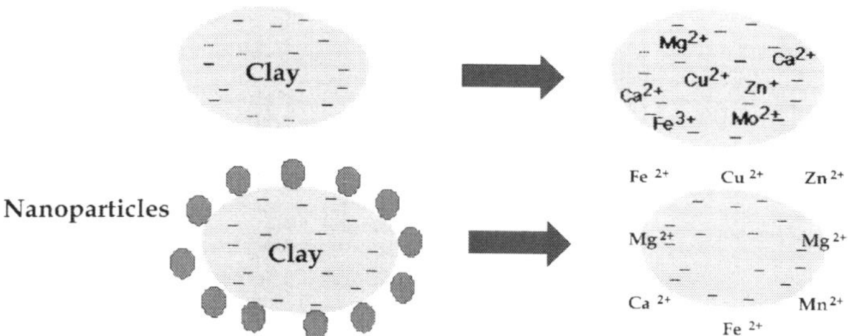

Fig. 21.1 : Nanoparticles catalyze soil fertility

One of the major roles for nanotechnology-enabled devices will be the increased use of autonomous sensors linked into a GPS system for real-time monitoring. These nanosensors could be distributed throughout the field where they can monitor soil conditions and crop growth. Wireless sensors are already being used in certain parts of the USA and Australia.

For example, one of the Californian vineyards, Pickberry, in Sonoma County has installed wifi systems with the help of the IT company, Accenture.9 The initial cost of setting up such a system is justified by the fact that it enables the best grapes to be grown which in turn produce finer wines, which command a premium price. The use of such wireless networks is of course not restricted to vineyards, for exampie Forbes Magazine has reported that small nanosensors are being used by Honeywell (a technology R&D company with global branches) to monitor grocery stores in Minnesota.10 This technology enables shop keepers to identify food items which have passed their expiry date and also reminds them to issue a new purchase order. The global market for wireless sensors is predicted to be 7 billion USD by 2010.11 The union of biotechnology and nanotechnology in sensors will create equipment of increased sensitivity, allowing an earlier response to environmental changes. For example:

- Nanosensors utilising carbon nanotubes12 or nano-cantilevers13 are small enough to trap and measure individual proteins or even small molecules.
- Nanoparticles or nanosurfaces can be engineered to trigger an electrical or chemical signal in the presence of a contaminant such as bacteria.
- Other nanosensors work by triggering an enzymatic reaction or by using nanoengineered branching molecules called dendrimers as probes to bind to target chemicals and proteins.14 Ultimately, precision farming, with the help of smart sensors, will allow enhanced productivity in agriculture by providing accurate information, thus helping farmers to make better decisions.

Smart Delivery Systems

The use of pesticides increased in the second half of the 20th century with DDT becoming one of the most effective and widespread throughout the world. However, many of these pesticides, including DDT were later found to be highly toxic, affecting human and animal health and as a result whole ecosystems. As a consequence they were banned. To maintain crop yields, Integrated Pest Management systems, which mix traditional methods of crop rotation with biological pest control methods, are becoming popular and implemented in many countries, such as Tunisia and India.

In the future, nanoscale devices with novel properties could be used to make agricultural systems "smart". For example, devices could be used to identify plant health issues before these become visible to the farmer. Such devices may be capable of responding to different situations by taking appropriate remedial action. If not, they will alert the farmer to the problem. In this way, smart devices will act as both a preventive and an early warning system. Such devices could be used to deliver chemicals in a controlled and targeted manner in the same way as nanomedicine has implications for drug delivery in humans.

Nanomedicine developments are now beginning to allow us to treat different diseases such as cancer in animals with high precision, and targeted delivery (to specific tissues and organs) has become highly successful.

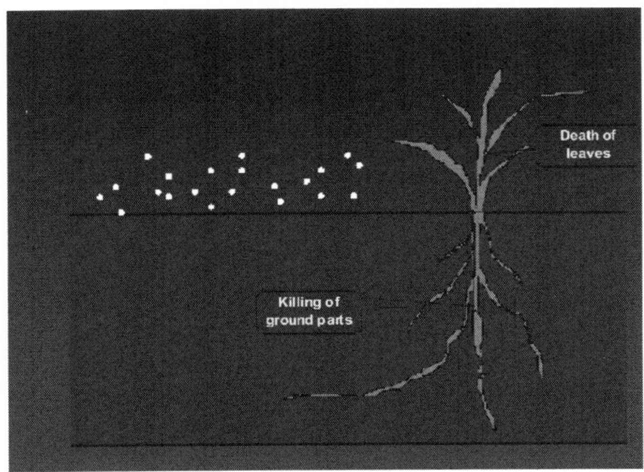

Fig. 21.2 : New herbicide targets domains

Technologies such as encapsulation and controlled release methods, have revolutionised the use of pesticides and herbicides. Many companies make formulations which contain nanoparticles within the 100-250 nm size range that are able to dissolve in water more effectively than existing ones (thus increasing their activity). Other companies employ suspensions of nanoscale particles (nanoemulsions), which can be either water or oil-based and contain uniform suspensions of pesticidal or herbicidal nanoparticles in the range of 200-400 nm. These can be easily incorporated in various media such as gels, creams, liquids etc, and have multiple applications for preventative measures, treatment or preservation of the harvested product.

One of the world's largest agrochemical corporations, Syngenta, is using nanoemulsions in its pesticide products. One of its successful growth regulating products is the Primo MAXX® plant growth regulator, which if applied prior to the onset of stress such as heat, drought,disease or traffic can strengthen the physical structure of turfgrass, and allow it to withstand ongoing stresses throughout the growing season.Another encapsulated product from Syngenta delivers a broad control spectrum on primary and secondary insect pests of cotton, rice, peanuts and soybeans. Marketed under the name Karate®

ZEON this is a quick release microencapsulated product containing the active compound lambda-cyhalothrin (a synthetic insecticide based on the structure of natural pyrethrins) which breaks open on contact with leaves. In contrast, the encapsulated product "gutbuster" only breaks open to release its contents when it comes into contact with alkaline environments, such as the stomach of certain insects. In other areas, scientists are working on various technologies to make fertiliser and pesticide delivery systems which can respond to environmental changes. The ultimate aim is to tailor these products in such a way that they will release their cargo in a controlled manner (slowly or quickly) in response to different signals e.g. magnetic fields, heat, ultrasound, moisture.

Protein binds with herbicide molecules

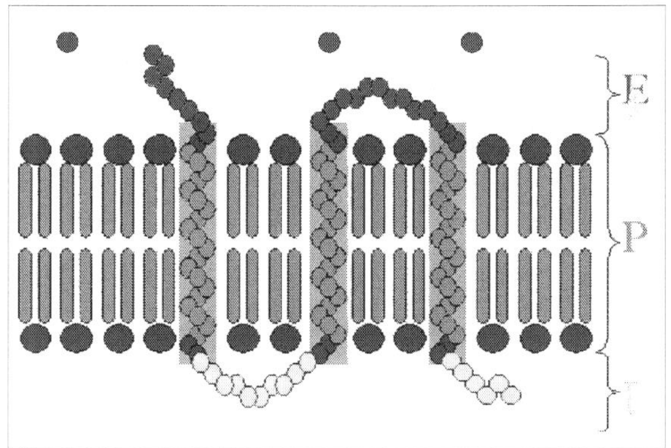

Transmembrane receptor:
E=extracellular space; I=intracellular space; P=plasma membrane

Fig. 21.3 : Slow Release Nano-fertilizers

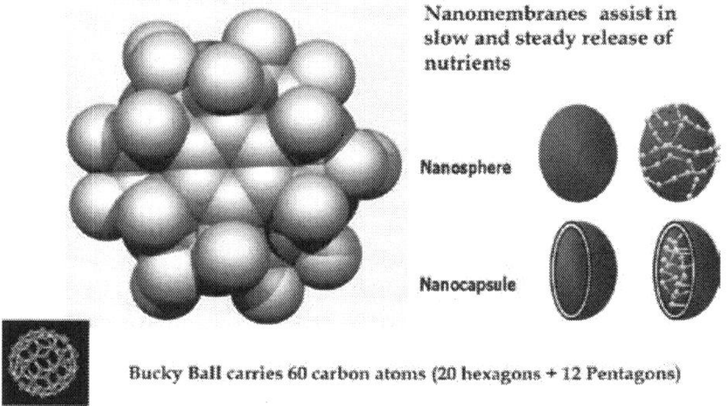

Fig. 21.4 : Slow Release Nano-fertilizers

New research also aims to make plants use water, pesticides and fertilizers more efficiently, to reduce pollution and to make agriculture more environmentally friendly. Smaller companies are forming alliances with major players such as LG, BASF, Honeywell, Bayer, Mitsubishi, and DuPont to make complete plant health monitoring systems in the next 10 years using nanotechnologies.

Other Developments in the Agricultural Sector due to Nanotechnology

Agriculture is the backbone of most developing countries, with more than 60% of the population reliant on it for their livelihood. As well as developing improved systems for monitoring environmental conditions and delivering nutrients or pesticides as appropriate, nanotechnology can improve our understanding of the biology of different crops and thus potentially enhance yields or nutritional values. In addition, it can offer routes to added value crops or environmental remediation.

Particle farming is one such example, which yields nanoparticles for industrial use by growing plants in defined soils. For example, research has shown that alfalfa plants grown in gold rich soil, absorb gold nanoparticles through their roots and accumulate these in their tissues. The gold nanoparticles can be mechanically separated from the plant tissue following harvest. Nanotechnology can also be used to clean ground water. The US company Argonide is using 2 nm diameter aluminium oxide nanofibres (NanoCeram) as a water purifier. Filters made from these fibres can remove viruses, bacteria and protozoan cysts from water.Similar projects are taking place elsewhere, particularly in developing countries such as India and South Africa. The German chemical group BASF's future business fund has devoted a significant proportion of its 105 million USD nanotechnology research fund to water purification techniques. The French utility company Generale des Eaux has also developed its own Nanofiltration technology in collaboration with the Dow Chemical subsidiary Filmtec.

Ondeo, the water unit of French conglomerate Suez, has meanwhile installed what it calls an ultrafiltration system, with holes of 0.1 microns in size, in one of its plants outside Paris. While some companies are working on water filtration, others such as Altairnano are following a purification approach. Altairnano's Nanocheck contains lanthanum nano particles that absorb phosphates from aqueous environments.

Applying these in ponds and swimming pools effectively removes available phosphates and as a result prevents the growth of algae. The company expects this product to benefit commercial fish ponds which spend huge amounts of money to remove algae. Research at Lehigh University in the US shows that an ultrafine, nanoscale powder made from iron can be used as an effective tool for cleaning up contaminated soil and groundwater- a trillion-dollar problem that encompasses more than 1000 still-untreated Superfund sites (uncontrolled or abandoned places where hazardous waste is located) in the United States, some 150,000 underground storage tank releases, and a huge number of landfills, abandoned mines, and industrial sites.The iron nanoparticles catalyse the oxidation and breakdown of organic contaminants such as trichloroethene, carbon tetrachloride, dioxins, and PCBs to simpler carbon compounds which are much less toxic.

This could pave the way for a nano-aquaculture, which would be beneficial for a large number of farmers across the world. Other research at the Centre for Biological and Environmental Nanotechnology (CBEN) has shown that nanoscale iron oxide particles are extremely effective at binding and removing arsenic from groundwater (something which affects the water supply of millions of people in the developing world, and for which there is no effective existing solution).

❑❑❑

Chapter-22

Nanomedicine

Nanomedicine is the medical application of nanotechnology. The approaches to nanomedicine range from the medical use of nanomaterials, to nanoelectronic biosensors, and even possible future applications of molecular nanotechnology. Current problems for nanomedicine involve understanding the issues related to toxicity and environmental impact of nanoscale materials.

Nanomedicine research is receiving funding from the US National Institute of Health. Of note is the funding in 2005 of a five-year plan to set up four nanomedicine centers. In April 2006, the journal Nature Materials estimated that 130 nanotech-based drugs and delivery systems were being developed worldwide.

Overview

Nanomedicine seeks to deliver a valuable set of research tools and clinically helpful devices in the near future. The National Nanotechnology Initiative expects new commercial applications in the pharmaceutical industry that may include advanced drug delivery systems, new therapies, and in vivo imaging. Neuro-electronic interfaces and other nanoelectronics-based sensors are another active goal of research. Further down the line, the speculative field of molecular nanotechnology believes that cell repair machines could revolutionize medicine and the medical field.

Nanomedicine is a large industry, with nanomedicine sales reaching 6.8 billion dollars in 2004, and with over 200 companies and 38 products worldwide, a minimum of 3.8 billion dollars in nanotechnology R&D is being invested every year. As the nanomedicine industry continues to grow, it is expected to have a significant impact on the economy.

Medical use of nanomaterials

Drug delivery

Nanomedical approaches to drug delivery center on developing nanoscale particles or molecules to improve the bioavailability of a drug. Bioavailability refers to the presence of drug molecules where they are needed in the body and where they will do the most good. Drug delivery focuses on maximizing bioavailability both at specific places in the body and over a period of time. This will be achieved by molecular targeting by nanoengineered devices. It is all about targeting the molecules and delivering drugs with cell precision. More than $65 billion are wasted each year due to poor bioavailability. *In vivo* imaging is another area where tools and devices are being developed. Using nanoparticle contrast agents, images such as ultrasound and MRI have a favorable distribution and improved contrast. The new methods of nanoengineered materials that are being developed might be effective in treating illnesses and diseases such as cancer. What nanoscientists will be able to achieve in the future is beyond current imagination. This will be accomplished by self assembled biocompatible nanodevices that will detect, evaluate, treat and report to the clinical doctor automatically.

Drug delivery systems, lipid- or polymer-based nanoparticles, can be designed to improve the pharmacological and therapeutic properties of drugs. The strength of drug delivery systems is their ability to alter the pharmacokinetics and biodistribution of the drug. Nanoparticles have unusual properties that can be used to improve drug delivery. Where larger particles would have been cleared from the body, cells take up these nanoparticles because of their size. Complex drug delivery mechanisms are being developed, including the ability to get drugs through cell membranes and into cell cytoplasm. Efficiency is important because many diseases depend upon processes within the cell and can only be impeded by drugs that make their way into the cell. Triggered response is one way for drug molecules to be used more efficiently. Drugs are placed in the body and only activate on encountering a particular signal. For example, a drug with poor solubility will be replaced by a drug delivery system where both hydrophilic and hydrophobic environments exist, improving the solubility. Also, a drug may cause tissue damage, but with drug delivery, regulated drug release can eliminate the problem. If a drug is cleared too quickly from the body, this could force a patient to use high doses, but with drug delivery systems clearance can be reduced by altering the pharmacokinetics of the drug. Poor biodistribution is a problem that can affect normal tissues through widespread distribution, but the particulates from drug delivery systems lower the volume of distribution and reduce the effect on non-target tissue. Potential nanodrugs will work by very specific and well-understood mechanisms; one of the major impacts of nanotechnology and nanoscience will be in leading development of completely new drugs with more useful behavior and less side effects.

Protein and Peptide Delivery

Protein and peptides exert multiple biological actions in human body and they have been identified as showing great promise for treatment of various diseases and disorders. These macromolecules are called biopharmaceuticals. Targetted and/or controlled delivery of these biopharmaceuticals using nanomaterials like nanoparticles and Dendrimers is an emerging field called nanobiopharmaceutics, and these products are called nanobiopharmaceuticals.

The small size of nanoparticles endows them with properties that can be very useful in oncology, particularly in imaging. Quantum dots (nanoparticles with quantum confinement properties, such as size-tunable light emission), when used in conjunction with MRI (magnetic resonance imaging), can produce exceptional images of tumor sites. These nanoparticles are much brighter than organic dyes and only need one light source for excitation. This means that the use of fluorescent quantum dots could produce a higher contrast image and at a lower cost than today's organic dyes used as contrast media. The downside, however, is that quantum dots are usually made of quite toxic elements.

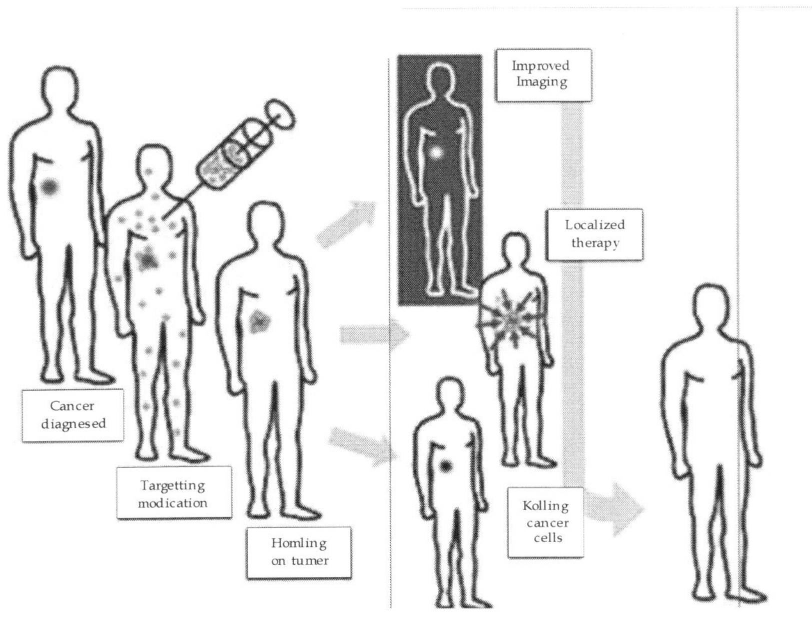

Fig. 22.1 : Molecular Imaging & Therapy

Another nanoproperty, high surface area to volume ratio, allows many functional groups to be attached to a nanoparticle, which can seek out and bind to certain tumor cells. Additionally, the small size of nanoparticles (10 to 100 nanometers), allows them to preferentially accumulate at tumor sites (because tumors lack an effective lymphatic drainage system). A very exciting research question is how to make these imaging nanoparticles do more things for cancer. For instance, is it possible to manufacture multifunctional nanoparticles that would detect, image, and then proceed to treat a tumor? This question is under vigorous investigation; the answer to which could shape the future of cancer treatment.A promising new cancer treatment that may one day replace radiation and chemotherapy is edging closer to human trials. Kanzius RF therapy attaches microscopic nanoparticles to cancer cells and then "cooks" tumors inside the body with radio waves that heat only the nanoparticles and the adjacent (cancerous) cells.

Sensor test chips containing thousands of nanowires, able to detect proteins and other biomarkers left behind by cancer cells, could enable the detection and diagnosis of cancer in the early stages from a few drops of a patient's blood.

The basic point to use drug delivery is based upon three facts: a) efficient encapsulation of the drugs, b) successful delivery of said drugs to the targeted region of the body, and c) successful release of that drug there.

Researchers at Rice University under Prof. Jennifer West, have demonstrated the use of 120 nm diameter nanoshells coated with gold to kill cancer tumors in mice. The nanoshells can be targeted to bond to cancerous cells by conjugating antibodies or peptides to the nanoshell surface. By irradiating the area of the tumor with an infrared laser, which passes through flesh without heating it, the gold is heated sufficiently to cause death to the cancer cells.

Additionally, John Kanzius has invented a radio machine which uses a combination of radio waves and carbon or gold nanoparticles to destroy cancer cells.

Nanoparticles of cadmium selenide (quantum dots) glow when exposed to ultraviolet light. When injected, they seep into cancer tumors. The surgeon can see the glowing tumor, and use it as a guide for more accurate tumor removal.

One scientist, University of Michigan's James Baker, believes he has discovered a highly efficient and successful way of delivering cancer-treatment drugs that is less harmful to the surrounding body. Baker has developed a nanotechnology that can locate and then eliminate cancerous cells. He looks at a molecule called a dendrimer. This molecule has over one hundred hooks on it that allow it to attach to cells in the body for a variety of purposes. Baker then attaches folic-acid to a few of the hooks (folic-acid, being a vitamin, is received by cells in the body). Cancer cells have more vitamin receptors than normal cells, so Baker's vitamin-laden dendrimer will be absorbed by the cancer cell. To the rest of the hooks on the dendrimer, Baker places anti-cancer drugs that will be absorbed with the dendrimer into the cancer cell, thereby delivering the cancer drug to the cancer cell and nowhere else (Bullis 2006).

In photodynamic therapy, a particle is placed within the body and is illuminated with light from the outside. The light gets absorbed by the particle and if the particle is metal, energy from the light will heat the particle and surrounding tissue. Light may also be used to produce high energy oxygen molecules which will chemically react with and destroy most organic molecules that are next to them (like tumors). This therapy is appealing for many reasons. It does not leave a "toxic trail" of reactive molecules throughout the body (chemotherapy) because it is directed where only the light is shined and the particles exist. Photodynamic therapy has potential for a noninvasive procedure for dealing with diseases, growths, and tumors.

Surgery

At Rice University, a flesh welder is used to fuse two pieces of chicken meat into a single piece. The two pieces of chicken are placed together touching. A greenish liquid containing gold-coated nanoshells is dribbled along the seam. An infrared laser is traced along the seam, causing the two sides to weld together. This could solve the difficulties and blood leaks caused when the surgeon tries to restitch the arteries s/he has cut during a kidney or heart transplant. The flesh welder could meld the artery perfectly.

Visualization

Tracking movement can help determine how well drugs are being distributed or how substances are metabolized. It is difficult to track a small group of cells throughout the body, so scientists used to dye the cells. These dyes needed to be excited by light of a certain wavelength in order for them to light up. While different color dyes absorb different frequencies of light, there was a need for as many light sources as cells. A way around this problem is with luminescent tags. These tags are quantum dots attached to proteins that penetrate cell membranes. The dots can be random in size, can be made of bio-inert material, and they demonstrate the nanoscale property that color is size-dependent. As a result, sizes are selected so that the frequency of light used to make a group of quantum dots fluoresce is an even multiple of the frequency required to make another group incandesce. Then both groups can be lit with a single light source.

Nanoparticle targeting

It is greatly observed that nanoparticles are promising tools for the advancement of drug delivery, medical imaging, and as diagnostic sensors.[*who?*] However, the biodistribution of these nanoparticles is mostly unknown due to the difficulty in targeting specific organs in the body. Current research in the excretory systems of mice, however, shows the ability of gold composites to selectively target certain organs based on their size and charge. These composites are encapsulated by a dendrimer and assigned a specific charge and size. Positively-charged gold nanoparticles were found to enter the kidneys while negatively-charged gold nanoparticles remained in the liver and spleen. It is suggested that the positive surface charge of the nanoparticle decreases the rate of osponization of nanoparticles in the liver, thus affecting the excretory pathway. Even at a relatively small size of 5 nm, though, these particles can become compartmentalized in the peripheral tissues, and will therefore accumulate in the body over time. While advancement of research proves that targeting and distribution can be augmented by nanoparticles, the dangers of nanotoxicity become an important next step in further understanding of their medical uses.

Neuro-electronic interfaces

Neuro-electronic interfacing is a visionary goal dealing with the construction of nanodevices that will permit computers to be joined and linked to the nervous system. This idea requires the building of a molecular structure that will permit control and detection of nerve impulses by an external computer. The computers will be able to interpret, register, and respond to signals the body gives off when it feels sensations. The demand for such structures is huge because many diseases involve the decay of the nervous system (ALS and multiple sclerosis). Also, many injuries and accidents may impair the nervous system resulting in dysfunctional systems and paraplegia. If computers could control the nervous system through neuro-electronic interface, problems that impair the system could be controlled so that effects of diseases and injuries could be overcome. Two considerations must be made when selecting the power source for such applications. They are refuelable and nonrefuelable strategies. A refuelable strategy implies energy is refilled continuously or periodically with external sonic, chemical, tethered, magnetic, or electrical sources. A nonrefuelable strategy implies that all power is drawn from internal energy storage which would stop when all energy is drained.

One limitation to this innovation is the fact that electrical interference is a possibility. Electric fields, electromagnetic pulses (EMP), and stray fields from other *in vivo* electrical devices can all cause interference. Also, thick insulators are required to prevent electron leakage, and if high conductivity of the *in vivo* medium occurs there is a risk of sudden power loss and "shorting out." Finally, thick wires are also needed to conduct substantial power levels without overheating. Little practical progress has been made even though research is happening. The wiring of the structure is extremely difficult because they must be positioned precisely in the nervous system so that it is able to monitor and respond to nervous signals. The structures that will provide the interface must also be compatible with the body's immune system so that they will remain unaffected in the body for a long time. In addition, the structures must also sense ionic currents and be able to cause currents to flow backward. While the potential for these structures is amazing, there is no timetable for when they will be available.

Medical applications of molecular nanotechnology

Molecular nanotechnology is a speculative subfield of nanotechnology regarding the possibility of engineering molecular assemblers, machines which could re-order matter at a molecular or atomic scale. Molecular nanotechnology is highly theoretical, seeking to anticipate what inventions nanotechnology might yield and to propose an agenda for future inquiry. The proposed elements of molecular nanotechnology, such as molecular assemblers and nanorobots are far beyond current capabilities.

Nanorobots

The somewhat speculative claims about the possibility of using nanorobots in medicine, advocates say, would totally change the world of medicine once it is realized. Nanomedicine would make use of these nanorobots (e.g., Computational Genes), introduced into the body, to repair or detect damages and infections. According to Robert Freitas of the Institute for Molecular Manufacturing, a typical blood borne medical nanorobot would be between 0.5-3 micrometres in size, because that is the maximum size possible due to capillary passage requirement. Carbon would be the primary element used to build these nanorobots due to the inherent strength and other characteristics of some forms of carbon (diamond/fullerene composites), and nanorobots would be fabricated in desktop nanofactories specialized for this purpose.

Nanodevices could be observed at work inside the body using MRI, especially if their components were manufactured using mostly ^{13}C atoms rather than the natural ^{12}C isotope of carbon, since ^{13}C has a nonzero nuclear magnetic moment. Medical nanodevices would first be injected into a human body, and would then go to work in a specific organ or tissue mass. The doctor will monitor the progress, and make certain that the nanodevices have gotten to the correct target treatment region. The doctor will also be able to scan a section of the body, and actually see the nanodevices congregated neatly around their target (a tumor mass, etc.) so that he or she can be sure that the procedure was successful.

Cell repair machines

Using drugs and surgery, doctors can only encourage tissues to repair themselves. With molecular machines, there will be more direct repairs. Cell repair will utilize the same tasks that living systems already prove possible. Access to cells is possible because biologists can stick needles into cells without killing them. Thus, molecular machines are capable of entering the cell. Also, all specific biochemical interactions show that molecular systems can recognize other molecules by touch, build or rebuild every molecule in a cell, and can disassemble damaged molecules. Finally, cells that replicate prove that molecular systems can assemble every system found in a cell. Therefore, since nature has demonstrated the basic operations needed to perform molecular-level cell repair, in the future, nanomachine based systems will be built that are able to enter cells, sense differences from healthy ones and make modifications to the structure.

The possibilities of these cell repair machines are impressive. Comparable to the size of viruses or bacteria, their compact parts would allow them to be more complex. The early machines will be specialized. As they open and close cell membranes or travel through tissue and enter cells and viruses, machines will only be able to correct a single molecular disorder like DNA damage or enzyme deficiency. Later, cell repair machines will be programmed with more abilities with the help of advanced AI systems.

Nanocomputers will be needed to guide these machines. These computers will direct machines to examine, take apart, and rebuild damaged molecular structures. Repair machines will be able to repair whole cells by working structure by structure. Then by working cell by cell and tissue by tissue, whole organs can be repaired. Finally, by working organ by organ, health is restored to the body. Cells damaged to the point of inactivity can be repaired because of the ability of molecular machines to build cells from scratch. Therefore, cell repair machines will free medicine from reliance on self repair.

Nanonephrology

Nanonephrology is a branch of nano medicine and nanotechnology that deals with 1) the study of kidney protein structures at the atomic level; 2) nano-imaging approaches to study cellular processes in kidney cells; and 3) nano medical treatments that utilize nanoparticles and to treat various kidney diseases. The creation and use of materials and devices at the molecular and atomic levels that can be used for the diagnosis and therapy of renal diseases is also a part of Nanonephrology that will play a role in the management of patients with kidney disease in the future. Advances in Nanonephrology will be based on discoveries in the above areas that can provide nano-scale information on the cellular molecular machinery involved in normal kidney processes and in pathological states. By understanding the physical and chemical properties of proteins and other macromolecules at the atomic level in various cells in the kidney, novel therapeutic approaches can be designed to combat major renal diseases. The nano-scale artificial kidney is a goal that many physicians dream of. Nano-scale engineering advances will permit programmable and controllable nano-scale robots to execute curative and reconstructive procedures in the human kidney at the cellular and molecular levels. Designing nanostructures compatible with the kidney cells and that can safely operate in vivo is also a future goal. The ability to direct events in a controlled fashion at the cellular nano-level has the potential of significantly improving the lives of patients with kidney diseases.

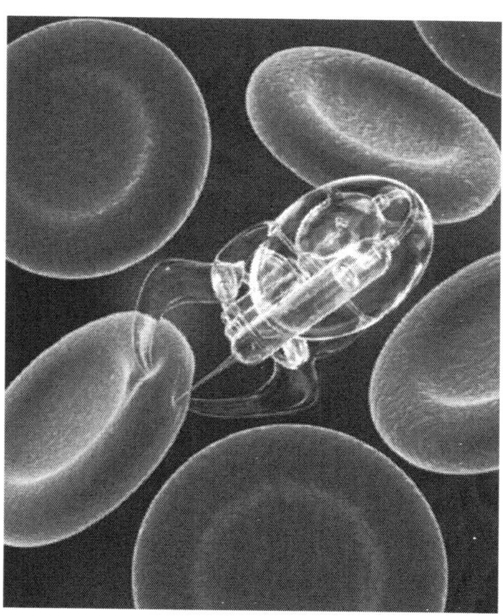

Fig. 22.2 : A Cell Repair Nanorobot

❑❑❑

Chapter-23

Nanotechnology in the Food Industry

The impact of nanotechnology in the food industry has become more apparent over the last few years with the organization of various conferences dedicated to the topic, initiation of consortia for better and safe food, along with increased coverage in the media. Several companies which were hesitant about revealing their research programmes in nanofood, have now gone public announcing plans to improve existing products and develop new ones to maintain market dominance. The types of application include: smart packaging, on demand preservatives, and interactive foods. Building on the concept of "on-demand" food, the idea of interactive food is to allow consumers to modify food depending on their own nutritional needs or tastes. The concept is that thousands of nanocapsules containing flavour or colour enhancers, or added nutritional elements (such as vitamins), would remain dormant in the food and only be released when triggered by the consumer.24 Most of the food giants including Nestle, Kraft, Heinz, and Unilever support specific research programmes to capture a share of the nanofood market in the next decade.

The definition of nanofood is that nanotechnology techniques or tools are used during cultivation, production, processing, or packaging of the food. It does not mean atomically modified food or food produced by nanomachines. Although there are ambitious thoughts of creating molecular food using nanomachines, this is unrealistic in the foreseeable future. Instead nanotechnologists are more optimistic about the potential to change the existing system of food processing and to ensure the safety of food products, creating a healthy food culture. They are also hopeful of enhancing the nutritional quality of food through selected additives and improvements to the way the body digests and absorbs food. Although some of these goals are further away, the food packaging industry already incorporates nanotechnology in products.

Packaging and Food Safety

Developing smart packaging to optimise product shelf-life has been the goal of many companies. Such packaging systems would be able to repair small holes/tears, respond to environmental conditions (e.g. temperature and moisture changes), and alert the customer if the food is contaminated. Nanotechnology can provide solutions for these, for example modifying the permeation behaviour of foils, increasing barrier properties (mechanical, thermal, chemical, and microbial), improving mechanical and heat-resistance properties, developing active antimicrobic and antifungal surfaces, and sensing as well as signalling microbiological and biochemical changes.

The financial outlook for nanotechnology enabled packaging looks buoyant. The current packaging market stands at 1.1 billion USD and is predicted to increase to 3.7 billion USD by 2010. Within this, the Smart Packaging industry is growing faster than predicted and is already showing signs of maturity. Research by the financial firm Frost and Sullivan, found that today's consumers demand much more from packaging in terms of protecting the quality, freshness and safety of foods, as well as convenience. They conclude that this is one of the main reasons behind the increased interest in innovative methods of packaging.

There are several organizations developing Smart Packaging systems. For example, Kraft foods, along with researchers at Rutgers University in the US, is developing an "electronic tongue" for inclusion in packaging. This consists of an array of nanosensors which are extremely sensitive to gases released by food as it spoils, causing the sensor strip to change colour as a result, giving a clear visible signal of whether the food is fresh or not.

Bayer Polymers has developed the Durethan KU2-2601 packaging film, which is lighter, stronger and more heat resistant than those currently on the market. The primary purpose of food packaging films is to prevent contents from drying out and to protect them from moisture and oxygen. The new film is known as a "hybrid system" that is enriched with an enormous number of silicate nanoparticles. These massively reduce the entrance of oxygen and other gases, and the exit of moisture, thus preventing food from spoiling.

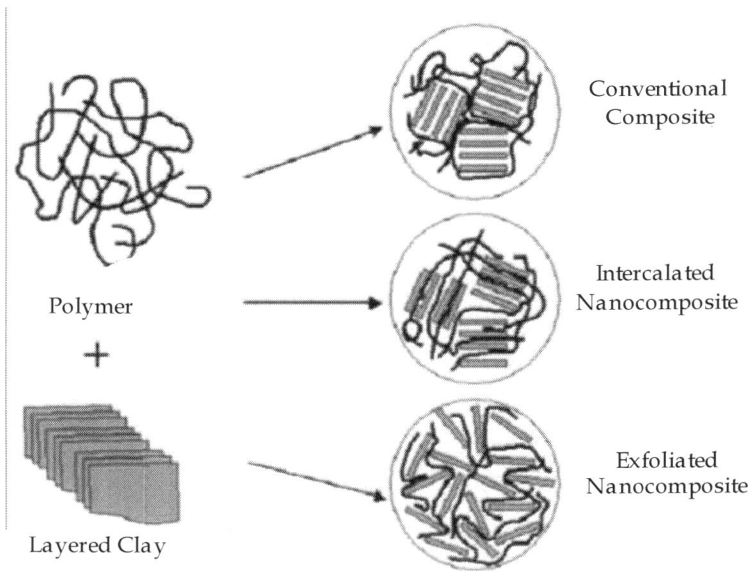

Fig 23.1 : Polymer Nanocomposites

Breweries would ideally use plastic bottles to ship beer, as these are lighter than glass and cheaper than metal cans. However, alcohol in beer reacts with the plastic used for the bottles, severely shortening shelf-life. Voridan, in association with Nanocor, has developed a nanocomposite containing clay nanoparticles, called Imperm. The resultant bottle is both lighter and stronger than glass and is less likely to shatter. The nanocomposite structure minimises loss of carbon dioxide from the beer and the ingress of oxygen to the bottle, keeping the beer fresher and giving it up to a six-month shelf life.

The technology has been adopted by several companies including the Miller Brewing Co.. Honeywell Specialty Polymers, has also successfully engineered plastic beer bottles that incorporate nanocomposites giving an extended shelf life (up to 26 weeks). The "Aegis" nylon 6 is the barrier layer in this 3-layered construction and has been used since late 2003 in the 1.6-litre Hite Pitcher beer bottle from Hite Brewery Co. in South Korea.

In a different strategy, Kodak is developing antimicrobial films that have the ability to absorb oxygen from the contents of the package, thus impeding food deterioration.

Other organizations are looking at ways in which nanotechnology can offer improvements in sensitivity or ease by which contamination of food is detected. For example, Agro Micron has developed the Nano Bioluminescence Detection Spray which contains a luminescent protein that has been engineered to bind to the surface of microbes such as Salmonella and E. coli. When bound, it emits a visible glow, thus allowing easy detection of contaminated food or beverages. The more intense the glow is, the higher the bacterial contamination. The company aims to market the product under the name BioMark and is currently designing new spray techniques to apply in ocean freight containerized shipping as well as to fight bioterrorism.

In a similar strategy to ensure food safety, EU researchers in the Good Food Project have developed a portable nanosensor to detect chemicals, pathogens and toxins in food.This circumvents the need to send samples to laboratories (which is both costly and lengthy), allowing food to be analysed for safety and quality at the farm, abattoir, during transport, processing or at the packaging plant. The project is also developing a device using DNA biochips to detect pathogens- a technique that could also be applied to determine the presence of different kinds of harmful bacteria in meat or fish, or fungi affecting fruit. The project also has plans to develop microarray sensors that can be used to identify pesticides on fruit and vegetables as well as those which will monitor environmental conditions at the farm. These have been coined "Good Food sensors".

The EU-funded BioFinger project, which has the aim of developing "versatile, inexpensive, and easy-to-use diagnostic tools for health, environmental and other applications", has found a different application in food analysis. The device uses cantilever technology, in which the tip of the cantilever is coated with chemicals allowing it to bend and resonate when it binds specific molecules (such as those on the surface of bacteria). The BioFinger device incorporates the cantilevers on a disposable microchip making it small and portable.32 The US military is developing super sensors to be used in times of terrorist attacks on food supplies. Current systems can take several days to confirm the presence of pathogens in food, however new nanotechnology enabled super sensors will be able to detect pathogens immediately. Such technology would have widespread applications in the food industry.

Researchers at the University of Bonn are developing dirt repellent coatings for packages using the lotus effect (water beads and runs off the surface of lotus leaves as a result of nanoscale wax pyramids which coat the leaves). Abattoirs and meat processing plants in particular could benefit from such technology. A research group at the University of Leeds in UK has determined that nanoparticles of magnesium oxide and zinc oxide are highly effective at destroying microorganisms. As these would be much cheaper to manufacture than silver nanoparticles, this could have tremendous applications in food packaging.

Nanotechnology has also found applications in monitoring and tagging of food items. Radio Frequency Identification (RFID) technology was developed by the military more than 50 years ago, but has now found its way to numerous applications from food monitoring in shops to improving supply chain efficiency. The technology, which consists of microprocessors and an antenna that can transmit data to a wireless receiver, can be used to monitor an item from the warehouse to the consumer's hands.34 Unlike bar codes, which need to be scanned manually and read individually, RFID tags do not require line-of-sight for reading and it is possible to automatically read hundreds of tags a second. Retailing chains like Wal-Mart, Home Depot, Metro group, and Tesco, have already tested this technology. The main drawback is the increased production costs due to silicon manufacturing. With the fusion of nanotechnology and electronics (nanotronics), these tags should become cheaper, easier to implement and more efficient.

A group of scientists from Northern European food industries have created a Nanofood consortium with the aim of fostering the applications of nanotechnology in the food industry in a responsible manner, to strengthen the effort to develop healthy and safe foods. The founding companies include Arla Foods, Danisco A/S, Aarhus United A/S, Danish Crown amba, Systematic Software Engineering A/S, and the

Interdisciplinary Nanoscience Centre (iNANO). With a mission to provide safe food to consumers, the consortium's priorities are: to develop sensors which can almost instantly reveal whether a food sample contains toxic compounds or bacteria; to develop anti-bacterial surfaces for machines involved in food production; to develop thinner, stronger and cheaper wrappings for food; and the creation of food with a healthier nutritional composition.35 A study by Denmark's Centre for Advanced Food Studies (LMC), an alliance of Danish institutions working in food sciences, has structured their priorities for the 7th Framework programme.

The six priority areas are:

- Basic understanding of food and animal feed for intelligent innovation
- Systems biology in food research
- Biological renewal in the food sector/biological production
- Technology development
- Nutrigenomics
- Consumer needs-driven innovation and food communication.

They believe that a focus on these areas will create a holistic and an interdisciplinary approach in food research and development in Europe. They are aiming to produce nanomaterials with functional properties, along with nanosensors and nanofluidic technology to be applied in food sciences. Other interests include the development of intelligent packaging materials, making it possible to monitor the condition of products during transportation or in display counters, and bio based packaging techniques.

Food Processing

In addition to packaging, nanotechnology is already making an impact on the development of functional or interactive foods, which respond to the body's requirements and can deliver nutrients more efficiently. Various research groups are also working to develop new "on demand" foods, which will remain dormant in the body and deliver nutrients to cells when needed. A key element in this sector is the development of nanocapsules that can be incorporated into food to deliver nutrients. Other developments in food processing include the addition of nanoparticles to existing foods to enable increased absorption of nutrients. One of the leading bakeries in Western Australia has been successful in incorporating nanocapsules containing tuna fish oil (a source of omega 3 fatty acids) in their top selling product "Tip-Top" Up bread. The microcapsules are designed to break open only when they have reached the stomach, thus avoiding the unpleasant taste of the fish oil.

The Israeli Company Nutralease, utilises Nano-sized Self-assembled Liquid Structures (NSSL) technology to deliver nutrients in nanosized particles to cells. The particles are expanded micelles (hollow spheres made from fats, with an aqueous interior) with a diameter of approximately 30 nm.38 The nutrients or "nutraceuticals" are contained within the aqueous interior. Nutraceuticals that have been incorporated in the carriers include lycopene, beta-carotene, lutein, phytosterols, CoQ10 and DHA/EPA. The Nutralease particles allow these compounds to enter the bloodstream from the gut more easily, thus increasing their bioavailability. The technology has already been adopted and marketed by Shemen Industries to deliver Canola Activa oil, which it claims reduces cholesterol intake into the body by 14%, by competing for bile solubilisation. This technology also has potential applications in the pharmaceutical industry.

A number of chemical companies are researching additives which are easily absorbed by the body and can increase product shelf life. Biodelivery Sciences International have developed nanocochleates, which are 50 nm coiled nanoparticles and can be used to deliver nutrients such as vitamins, lycopene, and omega fatty acids more efficiently to cells, without affecting the colour or taste of food.39 Kraft foods have established a consortium of research groups from 15 universities to look into the applications of nanotechnology to produce interactive foods. These will allow the consumer to choose between different flavours and colours. The consortium also has plans to develop smart foods which will release nutrients in response to deficiencies detected by nanosensors, and nanocapsules which will be ingested with food, but remain dormant until activated. All these new developments will make the concept of super foodstuffs a reality and these are expected to offer many different potential benefits including increased energy, improved cognitive functions, better immune function, and antiaging benefits.

Nanotechnology has already been used in the cosmetics industry to produce transparent creams. Royal BodyCare, a company utilizing nanotechnology in nutritional sciences, has marketed a new product called NanoCeuticals which is a colloid (or emulsion) of particles of less than 5 nm in diameter. The company claims the product will scavenge free radicals,increase hydration and balance the body's pH.40 The company has also developed NanoClustersTM, a nanosize powder combined with nutritional supplements. When consumed, it enhances the absorption of nutrients.

Food and Cosmetic Companies are working together to develop new mechanisms to deliver vitamins directly to the skin. For example, Nestlé, which has a 49% stake in L'Oréal, is developing transparent suncreams to deliver vitamin E directly to skin. The aim is to manufacture a cream which is absorbed by the skin and releases Vitamin E slowly, in addition to providing UV protection. Transparent UV-blocking creams are already on the market and L'Oréal expects the cream with added functionality to be marketed soon. Other competitors such as Estée Lauder are manufacturing anti-ageing formulations that make use of nanoparticles.

The US based Oilfresh Corporation has marketed a new nanoceramic product which reduces oil use in restaurants and fast food shops by half. As a result of its large surface area, the product prevents the oxidation and agglomeration of fats in deep fat fryers, thus extending the useful life span of the oil. An additional benefit is that oil heats up more quickly, reducing the energy required for cooking. Wageningen University in Netherlands has recently established a research centre which will focus its research on the application of nanotechnology in the food industry. The Wageningen BioNT (Bionanotechnology) Centre will concentrate on various topics including: sensing and diagnostics of food quality and safety; encapsulation and delivery of nutrients; micro- and nanodevices for physical and (bio) chemical processing; chemical biology; nanotoxicology; and consumer science and technology assessment.

The German company Aquanova has developed a new technology which combines two active substances for fat reduction and satiety into a single nano-carrier (micelles of average 30 nm diameter), an innovation said to be a new approach to intelligent weight management. Called NovaSOL Sustain, it uses CoQ1O to address fat reduction and alpha-lipoic acid for satiety. The NovaSol technology has also been used to create a vitamin E preparation that does not cloud liquids, called SoluE, and a vitamin C preparation called SoluC. The NovaSOL product can be used to introduce other dietary supplements as it protects contents from stomach acids.

In a different strategy, Unilever is developing low fat ice creams by decreasing the size of emulsion particles that give ice-cream its texture. By doing so it hopes to use up to 90% less of the emulsion and decrease fat content from 16% to about 1%. The Woodrow Wilson International Center for Scholars in the US has produced a consumer database of marketed nanotechnology and has so far identified more than 15 items which have a direct relation to the food industry. The list includes nanoceuticals developed by RBC Life Sciences and Canola Activa oil developed by Shemen Industries; the use of silver nanoparticles in refrigerators manufactured by LG Electricals, Samsung and Daewoo to inhibit bacterial growth and eliminate odours; All Spray For Life® which is manufactured by Health Plus International and uses a newly-designed pre-metered, non-aerosol Nanoceautical Delivery System (NDS) for transmucosal administration of dietary supplements, resulting in increased-bioavailability compared with gastrointestinal absorption.

❑❑❑

Chapter-24

Nanosensors

Nanosensors are any biological, chemical, or sugery sensory points used to convey information about nanoparticles to the macroscopic world. Their use mainly include various medicinal purposes and as gateways to building other nanoproducts, such as computer chips that work at the nanoscale and nanorobots. Presently, there are several ways proposed to make nanosensors, including Top-down and bottom-up design, top-down lithography, bottom-up assembly, and molecular self-assembly.

Predicted applications

Medicinal uses of nanosensors mainly revolve around the potential of nanosensors to accurately identify particular cells or places in the body in need. By measuring changes in volume, concentration, displacement and velocity, gravitational, electrical, and magnetic forces, pressure, or temperature of cells in a body, nanosensors may be able to distinguish between and recognize certain cells, most notably those of cancer, at the molecular level in order to deliver medicine or monitor development to specific places in the body. In addition, they may be able to detect macroscopic variations from outside the body and communicate these changes to other nanoproducts working within the body.

One example of nanosensors involves using the fluorescence properties of cadmium selenide quantum dots as sensors to uncover tumors within the body. By injecting a body with these quantum dots, a doctor could see where a tumor or cancer cell was by finding the injected quantum dots, an easy process because of their fluorescence. Developed nanosensor quantum dots would be specifically constructed to find only the particular cell for which the body was at risk. A downside to the cadmium selenide dots, however, is that they are highly toxic to the body. As a result, researchers are working on developing alternate dots made out of a different, less toxic material while still retaining some of the fluorescence properties.

In particular, they have been investigating the particular benefits of zinc sulfide quantum dots which, though they are not quite as fluorescent as cadmium selenide, can be augmented with other metals including manganese and various lanthanide elements. In addition, these newer quantum dots become more fluorescent when they bond to their target cells. (Quantum) Potential predicted functions may also include sensors used to detect specific DNA in order to recognize explicit genetic defects, especially for individuals at high-risk and implanted sensors that can automatically detect glucose levels for diabetic subjects more simply than current detectors. DNA can also serve as sacrificial layer for manufacturing CMOS IC, integrating a nanodevice with sensing capabilities. Therefore, using proteomic patterns and new hybrid materials, nanobiosensors can also be used to enable components configured into a hybrid semiconductor substrate as part of the circuit assembly. The development and miniaturization of nanobiosensors should provide interesting new opportunities.

Other projected products most commonly involve using nanosensors to build smaller integrated circuits, as well as incorporating them into various other commodities made using other forms of nanotechnology for use in a variety of situations including transportation, communication, improvements in structural integrity, and robotics. Nanosensors may also eventually be valuable as more accurate monitors of material states for use in systems where size and weight are constrained, such as in satellites and other aeronautic machines.

Existing nanosensors

Currently, the most common mass-produced functioning nanosensors exist in the biological world as natural receptors of outside stimulation. For instance, sense of smell, especially in animals in which it is particularly strong, such as dogs, functions using receptors that sense nanosized molecules. Certain plants, too, use nanosensors to detect sunlight; various fish use nanosensors to detect minuscule vibrations in the surrounding water; and many insects detect sex pheromones using nanosensors.

One of the first working examples of a synthetic nanosensor was built by researchers at the Georgia Institute of Technology in 1999[6]. It involved attaching a single particle onto the end of a carbon nanotube and measuring the vibrational frequency of the nanotube both with and without the particle. The discrepancy between the two frequencies allowed the researchers to measure the mass of the attached particle[1].

Chemical sensors, too, have been built using nanotubes to detect various properties of gaseous molecules. Carbon nanotubes have been used to sense ionization of gaseous molecules while nanotubes made out of titanium have been employed to detect atmospheric concentrations of hydrogen at the molecular level. Many of these involve a system by which nanosensors are built to have a specific pocket for another molecule. When that particular molecule, and only that specific molecule, fits into the nanosensor, and light is shone upon the nanosensor, it will reflect different wavelengths of light and, thus, be a different color.

Production methods

There are currently several hypothesized ways to produce nanosensors. Top-down lithography is the manner in which most integrated circuits are now made. It involves starting out with a larger block of some material and carving out the desired form. These carved out devices, notably put to use in specific microelectromechanical systems used as microsensors, generally only reach the micro size, but the most recent of these have begun to incorporate nanosized components.

Another way to produce nanosensors is through the bottom-up method, which involves assembling the sensors out of even more minuscule components, most likely individual atoms or molecules. This would involve moving atoms of a particular substance one by one into particular positions which, though it has been achieved in laboratory tests using tools such as atomic force microscopes, is still a significant difficulty, especially to do en masse, both for logistic reasons as well as economic ones. Most likely, this process would be used mainly for building starter molecules for self-assembling sensors.

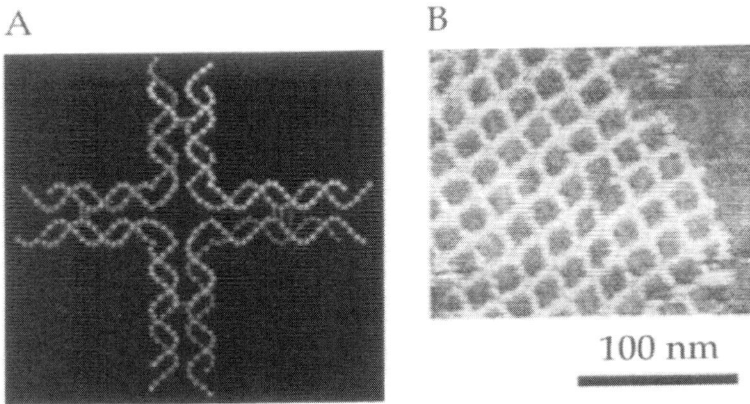

Fig. 24.1 : (A) An example of a DNA molecule used as a starter for larger self-assembly. (B) An atomic force microscope image of a self-assembled DNA nanogrid. Individual DNA tiles self-assemble into a highly ordered periodic two-dimensional DNA nanogrid.

The third way, which promises far faster results, involves self-assembly, or "growing" particular nanostructures to be used as sensors. This most often entails one of two types of assembly. The first involves using a piece of some previously created or naturally formed nanostructure and immersing it in free atoms of its own kind. After a given period, the structure, having an irregular surface that would make it prone to attracting more molecules as a continuation of its current pattern, would capture some of the free atoms and continue to form more of itself to make larger components of nanosensors.

The second type of self-assembly starts with an already complete set of components that would automatically assemble themselves into a finished product. Though this has been so far successful only in assembling computer chips at the micro size, researchers hope to eventually be able to do it at the nanometer size for multiple products, including nanosensors. Accurately being able to reproduce this effect for a desired sensor in a laboratory would imply that scientists could manufacture nanosensors much more quickly and potentially far more cheaply by letting numerous molecules assemble themselves with little or no outside influence, rather than having to manually assemble each sensor.

Economic impact

Though nanosensor technology is a relatively new field, global projections for sales of products incorporating nanosensors range from $0.6 billion to $2.7 billion in the next three to four years. They will likely be included in most modern circuitry used in advanced computing systems, since their potential to provide the link between other forms of nanotechnology and the macroscopic world allows developers to fully exploit the potential of nanotechnology to miniaturize computer chips while vastly expanding their storage potential.

First, however, nanosensor developers must overcome the present high costs of production in order to become worthwhile for implementation in consumer products. Additionally, nanosensor reliability is not yet suitable for widespread use, and, because of their scarcity, nanosensors have yet to be marketed and implemented outside of research facilities. Consequently, nanosensors have yet to be made compatible with most consumer technologies for which they have been projected to eventually enhance.

❑❑❑

Chapter-25

Impact of Nanotechnology in Energy and Environment

Over the past few decades, the fields of science and engineering have been seeking to develop new and improved types of energy technologies that have the capability of improving life all over the world. In order to make the next leap forward from the current generation of technology, scientists and engineers have been developing Energy Applications of Nanotechnology. Nanotechnology, a new field in science, is any technology that contains components smaller than 100 nanometers. For scale, a single virus particle is about 100 nanometers in width.

An important subfield of nanotechnology related to energy is nanofabrication. Nanofabrication is the process of designing and creating devices on the nanoscale. Creating devices smaller than 100 nanometers opens many doors for the development of new ways to capture, store, and transfer energy. The inherent level of control that nanofabrication could give scientists and engineers would be critical in providing the capability of solving many of the problems that the world is facing today related to the current generation of energy technologies.

People in the fields of science and engineering have already begun developing ways of utilizing nanotechnology for the development of consumer products. Benefits already observed from the design of these products are an increased efficiency of lighting and heating, increased electrical storage capacity, and a decrease in the amount of pollution from the use of energy. Benefits such as these make the investment of capital in the research and development of nanotechnology a top priority.

Consumer products

Recently, previously established and entirely new companies such as BetaBatt, Inc. and Oxane Materials are focusing on nanomaterials as a way to develop and improve upon older methods for the capture, transfer, and storage of energy for the development of consumer products.

ConsERV, a product developed by the Dais Analytic Corporation, uses nanoscale polymer membranes to increase the efficiency of heating and cooling systems and has already proven to be a lucrative design. The polymer membrane was specifically configured for this application by selectively engineering the size of the pores in the membrane to prevent air from passing, while allowing moisture to pass through the membrane. Polymer membranes can be designed to selectively allow particles of one size and shape to pass through while preventing others of different dimensions. This makes for a powerful tool that can be used in consumer products from biological weapons protection to industrial chemical separations.

A New York based company called Applied NanoWorks, Inc. has been developing a consumer product that utilizes LED technology to generate light. Light-emitting diodes or LEDs, use only about 10% of the energy that a typical incandescent or fluorescent light bulb use and typically lasts much longer, which makes them a viable alternative to traditional light bulbs. While LEDs have been around for decades, this company and others like it have been developing a special variant of LED called the white LED. White LEDs consist of semi-conducting organic layers that are only about 100 nanometers in distance from each other and are placed between two electrodes, which create an anode, and a cathode. When voltage is applied to the system, light is generated when electricity passes through the two organic layers. This is called electroluminescence. The semiconductor properties of the organic layers are what allow for the minimal amount of energy necessary to generate light. In traditional light bulbs, a metal filament is used to generate light when electricity is run through the filament. Using metal generates a great deal of heat and therefore lowers efficiency.

Research for longer lasting batteries has been an ongoing process for years. Researchers have now begun to utilize nanotechnology for battery technology. mPhase Technologies in conglomeration with Rutgers University and Bell Laboratories have utilized nanomaterials to alter the wetting behavior of the surface where the liquid in the battery lies to spread the liquid droplets over a greater area on the surface and therefore have greater control over the movement of the droplets. This gives more control to the designer of the battery. This control prevents reactions in the battery by separating the electrolytic liquid from the anode and the cathode when the battery is not in use and joining them when the battery is in need of use.

Thermal applications also are a future applications of nanothechonlogy creating low cost system of heating, ventilation, and air conditioning, changing molecular structure for better management of temperature

Economic benefits

The relatively recent shift toward using nanotechnology with respect to the capture, transfer, and storage of energy has and will continue to have many positive economic impacts on society. The control of materials that nanotechnology offers to scientists and engineers of consumer products is one of the most important aspects of nanotechnology. This allows for an improved efficiency of products across the board.

A major issue with current energy generation is the loss of efficiency from the generation of heat as a by-product of the process. A common example of this is the heat generated by the internal combustion engine. The internal combustion engine loses about 64% of the energy from gasoline as heat and an improvement of this alone could have a significant economic impact. However, improving the internal combustion engine in this respect has proven to be extremely difficult without sacrificing performance. Improving the efficiency of fuel cells through the use of nanotechnology appears to be more plausible by using molecularly tailored catalysts, polymer membranes, and improved fuel storage.

In order for a fuel cell to operate, particularly of the hydrogen variant, a noble-metal catalyst (usually platinum, which is very expensive) is needed to separate the electrons from the protons of the hydrogen atoms. However, catalysts of this type are extremely sensitive to carbon monoxide reactions. In order to combat this, alcohols or hydrocarbons compounds are used to lower the carbon monoxide concentration in the system. This adds an additional cost to the device. Using nanotechnology, catalysts can be designed through nanofabrication that are much more resistant to carbon monoxide reactions, which improves the efficiency of the process and may be designed with cheaper materials to additionally lower costs.

Fuel cells that are currently designed for transportation need rapid start-up periods for the practicality of consumer use. This process puts a lot of strain on the traditional polymer electrolyte membranes, which decreases the life of the membrane requiring frequent replacement. Using nanotechnology, engineers have the ability to create a much more durable polymer membrane, which addresses this problem. Nanoscale polymer membranes are also much more efficient in ionic conductivity. This improves the efficiency of the system and decreases the time between replacements, which lowers costs.

Another problem with contemporary fuel cells is the storage of the fuel. In the case of hydrogen fuel cells, storing the hydrogen in gaseous rather than liquid form improves the efficiency by 5%. However, the materials that we currently have available to us significantly limit fuel storage due to low stress tolerance and costs. Scientists have come up with an answer to this by using a nanoporous styrene material (which is a relatively inexpensive material) that when super-cooled to around -196°C, naturally holds on to hydrogen atoms and when heated again releases the hydrogen for use.

Environmental implications of nanotechnology

The environmental implications of nanotechnology are the possible effects that the use of nanotechnological materials and devices will have on the environment. As nanotechnology is an emerging field, there is great debate regarding to what extent industrial and commercial use of nanomaterials will affect organisms and ecosystems.

Nanotechnology's environmental implications can be split into two aspects: the potential for nanotechnologcal innovations to help improve the environment, and the possibly novel type of pollution that nanotechnological materials might cause if released into the envrionment.

Nanopollution is a generic name for all waste generated by nanodevices or during the nanomaterials manufacturing process. This kind of waste may be very dangerous because of its size. It can float in the air and might easily penetrate animal and plant cells causing unknown effects. Most human-made nanoparticles do not appear in nature, so living organisms may not have appropriate means to deal with nanowaste. It is probably[*who?*] one great challenge to nanotechnology: how to deal with its nanopollutants and nanowaste.

Environmental assessment is justified as nanoparticles present novel (new) environmental impacts. Scrinis[1] raises concerns about nano-pollution, and argues that it is not currently possible to "precisely predict or control the ecological impacts of the release of these nano-products into the environment." Ecotoxicological impacts of nanoparticles and the potential for bioaccumulation in plants and microorganisms remain under-researched. The capacity for nanoparticles to function as a transport mechanism also raises concern about the transport of heavy metals and other environmental contaminants. A May 2007 Report to the UK Department for Environment, Food and Rural Affairs noted concerns about the toxicological impacts of nanoparticles in relation to both hazard and exposure. The report recommended comprehensive toxicological testing and independent performance tests of fuel additives.

Not enough data exists to know for sure if nanoparticles could have undesirable effects on the environment. Two areas are relevant here: (1) In free form nanoparticles can be released in the air or water during production (or production accidents) or as waste by-product of production, and ultimately accumulate in the soil, water or plant life. (2) In fixed form, where they are part of a manufactured substance or product, they will ultimately have to be recycled or disposed of as waste. It is not known yet whether certain nanoparticles will constitute a completely new class of non-biodegradable pollutant. In case they do, it is not known how such pollutants could be removed from air or water because most traditional filters are not suitable for such tasks (their pores are too big to catch nanoparticles). Of the US$710 million spent in 2002 by the U.S. government on nanotechnology research, only $500,000 was spent on environmental impact assessments. Risks identified by Uskokovic (2007) include: self-replicating nanobots aggressively or through slowly rising supremacy wiping out the whole biosphere; further destabilising the already endangered diversity of the biosphere or extending the existing gap between the rich and poor.

Concerns have been raised about Silver Nano technology used by Samsung in a range of appliances such as washing machines and air purifiers.

Life cycle responsibility

To properly assess the health hazards of engineered nanoparticles the whole life cycle of these particles needs to be evaluated, including their fabrication, storage and distribution, application and potential abuse, and disposal. The impact on humans or the environment may vary at different stages of the life cycle.

The Royal Society report identified a risk of nanoparticles or nanotubes being released during disposal, destruction and recycling, and recommended that "manufacturers of products that fall under extended producer responsibility regimes such as end-of-life regulations publish procedures outlining how these materials will be managed to minimize possible human and environmental exposure" (p.xiii). Reflecting the challenges for ensuring responsible life cycle regulation, the Institute for Food and Agricultural Standards has proposed standards for nanotechnology research and development should be integrated across consumer, worker and environmental standards. They also propose that NGOs and other citizen groups play a meaningful role in the development of these standards.

Environmental benefits of nanotechnology

Energy

Nanotechnology could potentially have a great impact on clean energy production. Research is underway to use nanomaterials for purposes including more efficient solar cells, practical fuel cells, and environmentally-friendly batteries. The most advanced nanotechnology projects related to energy are: storage, conversion, manufacturing improvements by reducing materials and process rates, energy saving (by better thermal insulation for example), and enhanced renewable energy sources.

Current commercially available solar cells have low efficiencies of 15-20%. Research is ongoing to use nanowires and other nanostructured materials with the hope of to create cheaper and more efficient solar cells than are possible with conventional planar silicon solar cells. It is believed that these nanoelectronics-based devices will enable more efficient solar cells, and would have a great effect on satisfying global energy needs.

Another example for an environmentally friendly form of energy is the use of fuel cells powered by hydrogen. Probably the most prominent nanostructured material in fuel cells is the catalyst consisting of carbon supported noble metal particles with diameters of 1-5 nm. Suitable materials for hydrogen storage contain a large number of small nanosized pores.

Nanotechnology may also find applications in batteries. Because of the relatively low energy density of conventional batteries the operating time is limited and a replacement or recharging is needed, and the huge number of spent batteries represent a disposal problem. The use of nanomaterials may enable batteries with higher energy content or supercapacitors with a higher rate of recharging, which could be helpful for the battery disposal problem.

Water filtration and remediation

A strong influence of nanochemistry on waste-water treatment, air purification and energy storage devices is to be expected.

Mechanical or chemical methods can be used for effective filtration techniques. One class of filtration techniques is based on the use of membranes with suitable hole sizes, whereby the liquid is pressed through the membrane. Nanoporous membranes are suitable for a mechanical filtration with extremely small pores smaller than 10 nm ("nanofiltration") and may be composed of nanotubes. Nanofiltration is mainly used for the removal of ions or the separation of different fluids.

Magnetic nanoparticles offer an effective and reliable method to remove heavy metal contaminants from waste water by making use of magnetic separation techniques. Using nanoscale particles increases the efficiency to absorb the contaminants and is comparatively inexpensive compared to traditional precipitation and filtration methods.

Some water-treatment devices incorporating nanotechnology are already on the market, with more in development. Low-cost nanostructured separation membranes methods have been shown to be effective in producing potable water in a recent study.

Nanoscale iron particles have also shown potential as a detoxifying agent for cleaning environmental contaminents from brownfield sites.

❑❑❑

Chapter-26

Societal Implications of Nanotechnology

The societal implications of nanotechnology are the potential benefits and challenges that the introduction of novel nanotechnological devices and materials may hold for society and human interaction. The term is sometimes expanded to also include nanotechnology's health and environmental implications, but this article will only consider the social and political implications of nanotechnology.

As nanotechnology is an emerging field and most of its applications are still speculative, there is much debate about what positive and negative effects that nanotechnology might have.

Beyond the toxicity risks to human health and the environment which are associated with first-generation nanomaterials, nanotechnology has broader societal implications and poses broader social challenges. Social scientists have suggested that nanotechnology's social issues should be understood and assessed not simply as "downstream" risks or impacts. Rather, the challenges should be factored into "upstream" research and decision making in order to ensure technology development that meets social objectives.

Many social scientists and organizations in civil society suggest that technology assessment and governance should also involve public participation·

Some observers suggest that nanotechnology will build incrementally, as did the 18-19th century industrial revolution, until it gathers pace to drive a nanotechnological revolution that will radically reshape our economies, our labor markets, international trade, international relations, social structures, civil liberties, our relationship with the natural world and even what we understand to be human. Others suggest that it may be more accurate to describe change driven by nanotechnology as a "technological tsunami". Just like a tsunami, analysts warn that rapid nanotechnology-driven change will necessarily have profound disruptive impacts. As the APEC Center for Technology Foresight observes:

If nanotechnology is going to revolutionize manufacturing, health care, energy supply, communications and probably defense, then it will transform labour and the workplace, the medical system, the transportation and power infrastructures and the military. None of these latter will be changed without significant social disruption.

Those concerned with the negative implications of nanotechnology suggest that it will simply exacerbate problems stemming from existing socio-economic inequity and unequal distributions of power, creating greater inequities between rich and poor through an inevitable nano-divide (the gap between those who control the new nanotechnologies and those whose products, services or labour are displaced by them). Analysts suggest the possibility that nanotechnology has the potential to destabilize international relations through a nano arms race and the increased potential for bioweaponry; thus, providing the tools for ubiquitous surveillance with significant implications for civil liberties. Also, many critics believe it might break down the barriers between life and non-life through nanobiotechnology, redefining even what it means to be human.

Nanoethicists posit that such a transformative technology could exacerbate the divisions of rich and poor – the so-called "nano divide." However nanotechnology makes the production of technology, e.g. computers, cellular phones, health technology etcetera, cheaper and therefore accessible to the poor.

In fact, many of the most enthusiastic proponents of nanotechnology, such as transhumanists, see the nascent science as a mechanism to changing human nature itself - going beyond curing disease and enhancing human characteristics. Discussions on nanoethics have been hosted by the federal government, especially in the context of "converging technologies" - a catch-phrase used to refer to nano, biotech, information technology, and cognitive science.

Possible military applications

Societal risks from the use of nanotechnology have also been raised. On the instrumental level, these include the possibility of military applications of nanotechnology (for instance, as in implants and other means for soldier enhancement like those being developed at the Institute for Soldier Nanotechnologies at MIT [4]) as well as enhanced surveillance capabilities through nano-sensors.[9] There is also the possibility of nanotechnology being used to develop chemical weapons and because they will be able to develop the chemicals from the atom scale up, critics fear that chemical weapons developed from nano particles will be more dangerous than present chemical weapons.

Intellectual property issues

On the structural level, critics of nanotechnology point to a new world of ownership and corporate control opened up by nanotechnology. The claim is that, just as biotechnology's ability to manipulate genes went hand in hand with the patenting of life, so too nanotechnology's ability to manipulate molecules has led to the patenting of matter. The last few years has seen a gold rush to claim patents at the nanoscale. Over 800 nano-related patents were granted in 2003, and the numbers are increasing year to year. Corporations are already taking out broad-ranging patents on nanoscale discoveries and inventions. For example, two corporations, NEC and IBM, hold the basic patents on carbon nanotubes, one of the current cornerstones of nanotechnology. Carbon nanotubes have a wide range of uses, and look set to become crucial to several industries from electronics and computers, to strengthened materials to drug delivery and diagnostics. Carbon nanotubes are poised to become a major traded commodity with the potential to replace major conventional raw materials. However, as their use expands, anyone seeking to (legally) manufacture or sell carbon nanotubes, no matter what the application, must first buy a license from NEC or IBM.

The United State's essential facilities doctrine may be of importance as well as other anti-trust laws.

Potential benefits and risks for developing countries

Nanotechnologies may provide new solutions for the millions of people in developing countries who lack access to basic services, such as safe water, reliable energy, health care, and education. The United Nations has set Millennium Development Goals for meeting these needs. The 2004 UN Task Force on Science, Technology and Innovation noted that some of the advantages of nanotechnology include production using little labor, land, or maintenance, high productivity, low cost, and modest requirements for materials and energy.

Many developing countries, for example Costa Rica, Chile, Bangladesh, Thailand, and Malaysia, are investing considerable resources in research and development of nanotechnologies. Emerging economies such as Brazil, China, India and South Africa are spending millions of US dollars annually on R&D, and are rapidly increasing their scientific output as demonstrated by their increasing numbers of publications in peer-reviewed scientific publications.

Potential opportunities of nanotechnologies to help address critical international development priorities include improved water purification systems, energy systems, medicine and pharmaceuticals, food production and nutrition, and information and communications technologies. Nanotechnologies are already incorporated in products that are on the market. Other nanotechnologies are still in the research phase, while others are concepts that are years or decades away from development.

Applying nanotechnologies in developing countries raises similar questions about the environmental, health, and societal risks described in the previous section. Additional challenges have been raised regarding the linkages between nanotechnology and development.

Protection of the environment, human health and worker safety in developing countries often suffers from a combination of factors that can include but are not limited to lack of robust environmental, human health, and worker safety regulations; poorly or unenforced regulation which is linked to a lack of physical (e.g., equipment) and human capacity (i.e., properly trained regulatory staff). Often, these nations require assistance, particularly financial assistance, to develop the scientific and institutional capacity to adequately assess and manage risks, including the necessary infrastructure such as laboratories and technology for detection.

Very little is known about the risks and broader impacts of nanotechnology. At a time of great uncertainty over the impacts of nanotechnology it will be challenging for governments, companies, civil society organizations, and the general public in developing countries, as in developed countries, to make decisions about the governance of nanotechnology.

Companies, and to a lesser extent governments and universities, are receiving patents on nanotechnology. The rapid increase in patenting of nanotechnology is illustrated by the fact that in the US, there were 500 nanotechnology patent applications in 1998 and 1,300 in 2000. Some patents are very broadly defined, which has raised concern among some groups that the rush to patent could slow innovation and drive up costs of products, thus reducing the potential for innovations that could benefit low income populations in developing countries.

There is a clear link between commodities and poverty. Many least developed countries are dependent on a few commodities for employment, government revenue, and export earnings. Many applications of nanotechnology are being developed that could impact global demand for specific commodities. For instance, certain nanoscale materials could enhance the strength and durability of rubber, which might eventually lead to a decrease in demand for natural rubber. Other nanotechnology applications may result in increases in demand for certain commodities. For example, demand for titanium may increase as a result of new uses for nanoscale titanium oxides, such as titanium dioxide nanotubes that can be used to produce and store hydrogen for use as fuel. Various organizations have called for international dialogue on mechanisms that will allow developing countries to anticipate and proactively adjust to these changes.

In 2003, Meridian Institute began the Global Dialogue on Nanotechnology and the Poor: Opportunities and Risks (GDNP) to raise awareness of the opportunities and risks of nanotechnology for developing countries, close the gaps within and between sectors of society to catalyze actions that address specific opportunities and risks of nanotechnology for developing countries, and identify ways that science and technology can play an appropriate role in the development process. The GDNP has released several publicly accessible papers on nanotechnology and development, including "Nanotechnology and the Poor: Opportunities and Risks - Closing the Gaps Within and Between Sectors of Society"; "Nanotechnology, Water, and Development"; and "Overview and Comparison of Conventional and Nano-Based Water Treatment Technologies".

Social justice and civil liberties

Concerns are frequently raised that the claimed benefits of nanotechnology will not be evenly distributed, and that any benefits (including technical and/or economic) associated with nanotechnology will only reach affluent nations. The majority of nanotechnology research and development - and patents for nanomaterials and products - is concentrated in developed countries (including the United States, Japan, Germany, Canada and France). In addition, most patents related to nanotechnology are concentrated amongst few multinational corporations, including IBM, Micron Technologies, Advanced Micro Devices and Intel.[11] This has led to fears that it will be unlikely that developing countries will have access to the infrastructure, funding and human resources required to support nanotechnology research and development, and that this is likely to exacerbate such inequalities.

The agriculture and food industries demonstrate the concentration of nanotechnology related patents. Patents over seeds, plant material, animal and other agri-food techniques are already concentrated amongst a few corporations. This is anticipated to increase the cost of farming, by increasing farmers' input dependence. This may marginalize poorer farmers, including those living in developing countries.

Producers in developing countries could also be disadvantaged by the replacement of natural products (including rubber, cotton, coffee and tea) by developments in nanotechnology. These natural products are important export crops for developing countries, and many farmers' livelihoods depend on them. It has been argued that their substitution with industrial nano-products could negatively impact the economies of developing countries, that have traditionally relied on these export crops.

It is proposed that nanotechnology can only be effective in alleviating poverty and aid development "when adapted to social, cultural and local institutional contexts, and chosen and designed with the active participation by citizens right from the commencement point"

❑❑❑

Chapter-27

Nanotoxicology

Nanotoxicology is the study of the toxicity of nanomaterials. Because of quantum size effects and large surface area, nanomaterials have unique properties compared with their larger counterparts

Nanotoxicology is that branch of bionanoscience,which deals with the study and application of toxicity of nanomaterials. The nanomaterials, even when they are made of inert elements like gold, become very active at a nanometer range. Nanotoxicological studies are intended to determine whether and to what extent these may pose a threat to the environment and to human beings. For instance, Diesel nanoparticles have been found to damage the cardiovascular system in a mouse model.

Human health and safety

Calls for tighter regulation of nanotechnology have occurred alongside a growing debate related to the human health and safety risks associated with nanotechnology. The Royal Society identifies the potential for nanoparticles to penetrate the skin, and recommend that the use of nanoparticles in cosmetics be conditional upon a favorable assessment by the relevant European Commission safety advisory committee. Andrew Maynard also reports that 'certain nanoparticles may move easily into sensitive lung tissues after inhalation, and cause damage that can lead to chronic breathing problems'.

Carbon nanotubes - characterized by their microscopic size and incredible tensile strength - are frequently likened to asbestos, due to their needle-like fiber shape. In a recent study that introduced carbon nanotubes into the abdominal cavity of mice, results demonstrated that long thin carbon nanotubes showed the same effects as long thin asbestos fibers, raising concerns that exposure to carbon nanotubes may lead to mesothelioma (cancer of the lining of the lungs caused by exposure to asbestos). Given these risks, effective and rigorous regulation has been called for to determine if, and under what circumstances, carbon nanotubes are manufactured, as well as ensuring their safe handling and disposal.

The Woodrow Wilson Centre's Project on Emerging Technologies conclude that there is insufficient funding for human health and safety research, and as a result there is currently limited understanding of the human health and safety risks associated with nanotechnology. While the US National Nanotechnology Initiative reports that around four percent (about $40 million) is dedicated to risk related research and development, the Woodrow Wilson Centre estimate that only around $11 million is actually directed towards risk related research. They argued in 2007 that it would be necessary to increase funding to a minimum of $50 million in the following two years so as to fill the gaps in knowledge in these areas.

The potential for workplace exposure was highlighted by the 2004 Royal Society report which recommended a review of existing regulations to assess and control workplace exposure to nanoparticles and nanotubes. The report expressed particular concern for the inhalation of large quantities of nanoparticles by workers involved in the manufacturing process.

Stakeholders concerned by the lack of a regulatory framework to assess and control risks associated with the release of nanoparticles and nanotubes have drawn parallels with bovine spongiform encephalopathy ('mad cow's disease), thalidomide, genetically modified food),) nuclear energy, reproductive technologies, biotechnology, and asbestosis. In light of such concerns, the Canadian based ETC Group have called for a moratorium on nano-related research until comprehensive regulatory frameworks are developed that will ensure workplace safety.

Toxicology of nanoparticles

Nanotoxicology is a sub-specialty of particle toxicology. It addresses the toxicology of nanoparticles (particles <100 nm diameter) which appear to have some toxic effects that are unusual and not seen with larger particles. Nanoparticles can be divided into combustion-derived nanoparticles (like diesel soot), manufactured nanoparticles like carbon nanotubes and naturally occurring nanoparticles from volcanic eruptions, atmospheric chemistry etc. Typical nanoparticles that have been studied are titanium dioxide, alumina, zinc oxide, carbon black, and carbon nanotubes, and "nano-C_{60}". Nanoparticles seem to have some different properties from larger particles that are known to have pathogenic effects, like asbestos or quartz. These differences seem to be a result of their size. They have a larger surface area per unit mass, so that in some cases they may have more pro-inflammatory effects (in, for example, lung tissue). In addition, some nanoparticles seem to be able to translocate from their site of deposition to distant sites such as the blood and the brain. This has resulted in a sea-change in how particle toxicology is viewed- instead of being confined to the lungs, nanoparticle toxicologists study the brain, blood, liver, skin and gut. Nanotoxicology has revolutionised particle toxicology and rejuvenated it.

The smaller a particle is, the greater its surface area to volume ratio and the higher its chemical reactivity and biological activity. The greater chemical reactivity of nanomaterials results in increased production of reactive oxygen species (ROS), including free radicals. ROS production has been found in a diverse range of nanomaterials including carbon fullerenes, carbon nanotubes and nanoparticle metal oxides. ROS and free radical production is one of the primary mechanisms of nanoparticle toxicity; it may result in oxidative stress, inflammation, and consequent damage to proteins, membranes and DNA.

The extremely small size of nanomaterials also means that they much more readily gain entry into the human body than larger sized particles do. How these nanoparticles behave inside the body is a major question that needs to be resolved. The behavior of nanoparticles is a function of their size, shape and surface reactivity with the surrounding tissue. In principle, a large number of particles could overload the body's phagocytes, cells that ingest and destroy foreign matter, thereby triggering stress reactions that lead to inflammation and weaken the body's defense against other pathogens. In addition to questions about what happens if non-degradable or slowly degradable nanoparticles accumulate in bodily organs, another concern is their potential interaction with biological processes inside the body. Because of their large surface area, nanoparticles will, on exposure to tissue and fluids, immediately adsorb onto their surface some of the macromolecules they encounter. This may, for instance, affect the regulatory mechanisms of enzymes and other proteins.

Nanomaterials are able to cross biological membranes and access cells, tissues and organs that larger-sized particles normally cannot. Nanomaterials can gain access to the blood stream following inhalation or ingestion. At least some nanomaterials can penetrate the skin; even larger microparticles may penetrate skin when it is flexed. Broken skin is an ineffective particle barrier, suggesting that acne, eczema, shaving wounds or severe sunburn may accelerate skin uptake of nanomaterials. Then, once in the blood stream, nanomaterials can be transported around the body and be taken up by organs and tissues, including the brain, heart, liver, kidneys, spleen, bone marrow and nervous system. Nanomaterials have proved toxic to human tissue and cell cultures, resulting in increased oxidative stress, inflammatory cytokine production and cell death. Unlike larger particles, nanomaterials may be taken up by cell mitochondria and the cell nucleus. Studies demonstrate the potential for nanomaterials to cause DNA mutation and induce major structural damage to mitochondria, even resulting in cell death. Size is therefore a key factor in determining the potential toxicity of a particle. However it is not the only important factor.

Other properties of nanomaterials that influence toxicity include: chemical composition, shape, surface structure, surface charge, aggregation and solubility, and the presence or absence of functional groups of other chemicals. The large number of variables influencing toxicity means that it is difficult to generalise about health risks associated with exposure to nanomaterials - each new nanomaterial must be assessed individually and all material properties must be taken into account.

Since there is no authority to govern nanotech-based products, there are many products that could possibly be dangerous to humans.

Scientific research has indicated the potential for some nanomaterials to be toxic to humans or the environment. In March 2004 tests conducted by environmental toxicologist Eva Oberdörster, Ph.D. working with Southern Methodist University in Texas, found extensive brain damage to fish exposed to fullerenes for a period of just 48 hours at a relatively moderate dose of 0.5 parts per million (commensurate with levels of other kinds of pollution found in bays). The fish also exhibited changed gene markers in their livers, indicating their entire physiology was affected. In a concurrent test, the fullerenes killed water fleas, an important link in the marine food chain. The extremely small size of fabricated nanomaterials also means that they are much more readily taken up by living tissue than presently known toxins. Nanoparticles can be inhaled, swallowed, absorbed through skin and deliberately or accidentally injected during medical procedures. They might be accidentally or inadvertently released from materials implanted into living tissue.

Researcher Shosaku Kashiwada of the National Institute for Environmental Studies in Tsukuba, Japan, in a more recent study, intended to further investigate the effects of nanoparticles on soft-bodied organisms. His study allowed him to explore the distribution of water-suspended fluorescent nanoparticles throughout the eggs and adult bodies of a species of fish, known as the see-through medaka (Oryzias latipes). See-through medakas were used because of their small size, wide temperature and salinity tolerances, and short generation time. Moreover, small fish like the see-through medaka have been popular test subjects for human diseases and organogenesis for other reasons as well, including their transparent embryos, rapid embryo development, and the functional equivalence of their organs and tissue material to that of mammals. Because the see-through medakas have transparent bodies, analyzing the deposition of fluorescent nanoparticles throughout the body is quite simple. For his study, Dr. Kashiwada evaluated four aspects of nanoparticle accumulation. These included the overall accumulation and the size-dependent accumulation of nanoparticles by medaka eggs, the effects of salinity on the aggregation of nanoparticles in solution and on their accumulation by medaka eggs, and the distribution of nanoparticles in the blood and organs of adult medaka. It was also noted that nanoparticles were in fact taken up into the bloodstream and deposited throughout the body. In the medaka eggs, there was a high accumulation of nanoparticles in the yolk; most often bioavailibility was dependent on specific sizes of the particles. Adult samples of medaka had accumulated nanoparticles in the gills, intestine, brain, testis, liver, and bloodstream. One major result from this study was the fact that salinity may have a large influence on the bioavailibility and toxicity of nanoparticles to penetrate membranes and eventually kill the specimen.

As the use of nanomaterials increases worldwide, concerns for worker and user safety are mounting. To address such concerns, the Swedish Karolinska Institute conducted a study in which various nanoparticles were introduced to human lung epithelial cells. The results, released in 2008, showed that iron oxide nanoparticles caused little DNA damage and were non-toxic. Zinc oxide nanoparticles were slightly worse. Titanium dioxide caused only DNA damage. Carbon nanotubes caused DNA damage at low levels. Copper oxide was found to be the worst offender, and was the only nanomaterial identified by the researchers as a clear health risk.

Immunogenicity to nanoparticles:

Very little attention has been concentrated in the potential immunogenicity of nanostructures. Nanostructures can activate the immune system inducing inflammation, immune responses, allergy, or even affect to the immune cells in a deleterious or beneficial way (immunosuppression in autoinmmunity diseases, improving immune responses in vaccines). Many studies are needed in order to know the potencial deleterous or beneficial effects of nanostructures in the immune system. Comparing to the conventional pharmeceutical agents, nanostructures have a huge size and immune cells, especially phagocytic cells, recognize and try to destroy them.

In addition, standarization of toxicology tests between laboratories are needed. Díaz, B. et al from the University of Vigo (Spain) has shown (Small, 2008) that many different cell lines should be studied in order to know if a nanostructure induces toxicity, and human cells can intenalize aggregated nanoparticles. Moreover, it is important to take into account that many nanostructures aggregate in biological fluids, but groups manufacturing nanostructures do not care much about this matter. Many efforts of interdisciplinary groups are strongly needed in order to progress in this field.

No Fullerene toxicity reported

Nanoparticles can also be made of C_{60}, as is the case with almost any room temperature solid, and several groups have done this and studied toxicity of such particles. The results in the work of Oberdörster at Southern Methodist University, published in "Environmental Health Perspectives" in July 2004, in which questions were raised of potential cytotoxicity, has now been shown by several sources to be likely caused by the tetrahydrofuran used in preparing the 30 nm–100 nm particles of C_{60} used in the research. Isakovic, et al., 2006, who review this phenomenon, gives results showing that removal of THF from the C_{60} particles resulted in a loss of toxicity. Sayes, et al., 2007, also show that particles prepared as in Oberdorster caused no detectable inflammatory response when instilled intratracheally in rats after observation for 3 months, suggesting that even the particles prepared by Oberdorster do not exhibit markers of toxicity in mammalian models. This work used as a benchmark quartz particles, which did give an inflammatory response.

A comprehensive and recent review of work on fullerene toxicity is available in "Toxicity Studies of Fullerenes and Derivatives," a chapter from the book "Bio-applications of Nanoparticles". In this work, the authors review the work on fullerene toxicity beginning in the early 1990s to present, and conclude that the evidence gathered since the discovery of fullerenes overwhelmingly points to C_{60} being non-toxic. As is the case for toxicity profile with any chemical modification of a structural moiety, the authors suggest that individual molecules be assessed individually.

❑❑❑

Chapter-28

Nanotechnology Developments : An Overview

Nanotechnology, which deals with understanding and control of matter at dimension of roughly 100 nm and below, has a cross sectoral application and an interdisciplinary orientation. At this scale, the physical, chemical and biological properties of materials differ from the properties of individual atoms and molecules or bulk matter, which enable novel applications. Nanotechnology research and development is directed towards understanding and creating improved materials, devices and systems that exploit these properties as they are discovered and characterized

There are many applications of nanotechnology such as in the area of medicine, chemistry and environment, energy, agriculture, information and communication, heavy industry and consumer goods. The alleged potential of this technology has garnered the attention of both developed and developing countries across the globe. The US National Science Foundation (NSF) has listed it as one of six priority areas; it is one of the themes in the EU Framework Program for Research and Technological Development in Europe; and it has been the focus of research in countries worldwide.

Globally, investments have been made, nanotechnology programmes initiated and research and development has commenced. It is observed that US, Japan, and Germany dominate the current R&D effort in nanotechnology with the country focus largely based on their own expertise and needs. Globally, there has been an increase in expenditure by both governments and private companies in nanotechnology developments. Total global expenditure (public + private) in nanotechnology R&D in 2007 amounted to $13.5 billion, up 14% from in 2006. Expenditure by corporations in nanotechnology R&D in 2007 witnessed a 23% increase over 2006 to reach $6.6 billion, passing government spending for the first time.

At the commercial level, the impact of nanotechnology, is evident in three major industry sectors, viz., materials and manufacturing (coatings and composites for products like automobiles and buildings), electronics (displays and batteries) and health care and life sciences (pharmaceutical applications). According to the Woodrow Wilson International Center for Scholars' Project on Emerging Nanotechnologies (2009), there are more than 1000 company-identified nanotechnology products on the market, the majority being produced by companies based in USA.

The categorization of nanotechnology products indicates a concentration in the fields of health and fitness products (cosmetics, clothing, personal care and sporting equipment).The analysis of the product category suggests that nanotechnology mainly impacts the consumer goods industries. An estimate by Lux Research indicates that nanotechnology-derived revenues will attain 15% of projected global manufacturing output ($2.6 trillion) in 2014 as compared to 0.1% in 2006 ($50 billion).

Nanotechnology initiatives in India

The emergence of nanotechnology in India has witnessed the engagement of a diverse set of players, each with their own agenda and role. Nanotechnology in India is a government led initiative. Industry participation has very recently originated. Nanotechnology R&D barring a few exceptions is largely being ensued at public funded universities as well as research institutes.

Given the enabling nature of nanotechnology and ability to develop along with existing technologies, it has the potential to be utilised as a tool to address key development related challenges in diverse sectors like energy, water, agriculture, health, environment and the like. Enabling energy storage, production and conversion within renewable energy frameworks has been cited as the primary area where nanotechnology applications might aid developing countries. Nanotechnology interventions might be sought at specific junctures to improve quantity and quality of water and wastewater treatment systems.

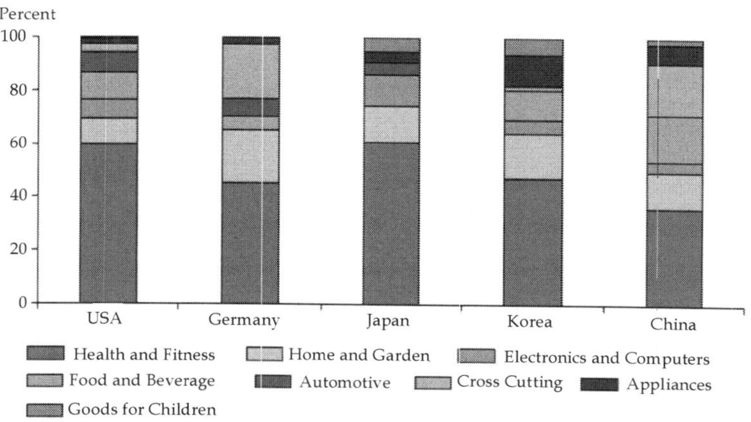

Country	Number of Products
USA	563
Germany	78
Japan	43
Korea	139
China	56

Fig. 28.1 : Percentage shares of different categories of nanotechnology products by country, October 2009, Source Woodrow Wilson International Centre for Scholars (www.nanotechproject. org)

Enhancement of agricultural productivity has been identified as a critical area of nanotechnology application nanotechnology for attaining the Millennium Development Goals. In light of the developments worldwide hailing nanotechnology as a technology with the potential of addressing a number of developing country needs, India has sought to promote nanotechnology applications in sectors that are likely to have a wide impact, and influence the course of future development in the country. Sectors such as health, energy and environment have received greater attention by various technology departments in the government (DST, DBT and SERC). Figure shows the distribution of sponsored projects from 2006 to 2008 across key sectors.

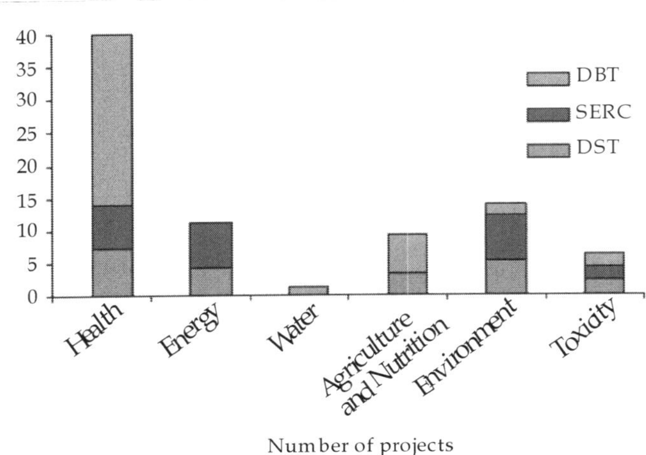

Fig. 28.2 : Sectorwise distribution of projects sponsored by DST, DBT and SERC – 2006–08
Source http://nanomission.gov.in/Reference: http://www.teriin.org/div/ST_BriefingPap.pdf

The efforts made by the government to promote nanotechnology

The government has been at the forefront of promoting nanotechnology industry in India through its three major funding agencies, namely Department of Science and Technology (DST, http://dst.gov.in), Department of Scientific and Industrial Research (DSIR, http://dsir.nic.in), and Department of Biotechnology (DBT http://dbtindia.nic.in).

DST launched the Nano Science and Technology Initiative (NSTI) in 2001 under the leadership of Prof. C. N. R. Rao. Its aim is to make India a major player in this sector, and provided a total of $15 million for nanotechnology over 5 years. The NSTI funded about 100 research projects, and provided funding for setting up 10 core groups in nano science, 6 centres of nanotechnology, and one of computational materials science at different institutions across India. DST remains by far the largest funding agency. Government spending in nanotechnology through all its funding agencies amounted to less than $20 million in 2003/04 out of total R&D expenditure of about $3.03 billion (0.8% of

GNP of India). According to government figures, the government has spent approximately $50 million over the past five years to promote R&D in the area of nanotechnology.

R&D activity in India is concentrated at the top 20 premier institutions like Indian Institute of Science, Jawaharlal Nehru Center for Advanced Scientific Research, various Indian Institute of Technology, and universities like Anna University, University of Delhi. Defense organizations like Defense Research and Development Organization (DRDO) are also active contributors to R&D output from the country. These organizations will be profiled in greater detail in these columns. These institutes focus on basic research and include following research areas: synthesis of nanomaterials and nanocomposites, thin films processing and characterization, nano-sensors, nano-biotechnology, and superconductivity, micro- and nano-fabrication.

Some institutes of India that offer nano- technology course are:

- Indian Institute of Science (Bangalore)
- Jawaharlal Nehru Center for Advanced Scientific Research (Bangalore)
- National Chemical Laboratory (Pune)
- National Physical Laboratory (Delhi)
- Defence Materials Store Research & Development Organization, (Kanpur)
- IITs (at Chennai, Mumbai, Delhi, Kanpur and Guwahati)
- Central Scientific Instruments Organization (Chandigarh)
- Amity Institute of Nano-technology (Noida)

What are the nanotechnology career prospects available in India?

With a professional degree in Nanotechnology under one's belt, one will get job opportunities in biotech companies among several other avenues. In a biotech company, one's work profile may include the fabrication of miniature systems and devices to be used in Nanomedicine or working on a nanoparticle-based molecular system for detecting biological warfare agents. One will also find employment opportunities in large pharmaceutical companies where one will be working on the delivery process of drugs or on the development of a new therapeutic drug.

There are a lot of opportunities available in the field of research in Nanotechnology; various research programmes in Nanotechnology are funded by the Government and universities across the country.

Nanotechnology is one field in which there are employment opportunities for engineers from all disciplines, The synthesis, functionalism characterizations, and optimization of materials are done by Chemical Engineers, which are a few among the variety of functions performed by them in the field of Nanotechnology, The development of non-material assembly processes, creation of test protocols and the preparation of reports are some of the other functions performed by Chemical Engineers. The design, production and testing of machines, instruments, controls, engines and mechanical, thermal or heat transfer systems is done by Mechanical Engineers. The designing of methods by virtue of which natural substances get changed into newer, stronger, and more resistant materials with unique properties is the work of Materials Engineers in the field of Nanotechnology.

Currently, there are around 500 MNC's in the market who offer nanotech products while another 150 odd institutions are involved in research work in Nanotechnology, The Government of India offers considerable support for the generation of awareness and promotion of Nanotechnology, Two schemes related to Nanotechnology-the Nanoscience and Technology Initiative launched by the Department of Science and Technology and the Science and Technology Initiatives in Nanotechnology-are being promoted by the Government of India.

Several other work opportunities are also available in Nanotechnology, apart from that of a scientist or an engineer, Business development and administration, legal areas, and sales and marketing are few other areas, where one can work in the field of Nanotechnology.With a professional degree in Nanotechnology under one's belt, one can work as a scientist, academician, biotechnologist, systems designer, research officer or product designer.

Nanotechnology impacts all major sectors like solar energy, aerospace, environment, telecommunications, computing, etc. Nanotechnology has been widely used in number of movies, television series and video games too. By the year 2015, Nanotechnology is estimated to have grown to a $1 trillion industry. It is expected to revolutionize a number of fields like medicine and military. It is the technology of the future.

❑❑❑

Glossary

ACE Paste: **A**tomspheric **C**arbon **E**xtractor. Harvests the greenhouse gases for Carbon, to be used for diamondoid fabrication. Larger than most pastebots, because it has to be collectible afterwards. A well-designed paste could harvest 100X or more its empty weight. ACE Paste may not be necessary, because large fixed installations might be more efficient.

Adensoine Triphosphate [ATP]: A chemical compound that functions as fuel for biomolecular nanotechnology having the formula, $C_{10}H_{16}N_5O_{13}P_3$.

Animat:This term was developed by Alan H. Goldstein. In his article I, Nanobot, he suggested that a new state of life be named after the contraction of the term "anima-materials" Ñ "animats". This artificial life form (most likely nano biotechnology based) must meet the following tests:

A = Devices that can survive and function in our ecosphere, for example inside human beings.

B = Devices that can derive energy from biological metabolism. Many nanomedical devices will be powered by the fuel available inside the human body. A common idea is to take our own glucose-oxidizing enzymes and use them as a fuel cell for the nanobiobot.

C = Devices capable of copying themselves by molecular self-assembly. Note that any information necessary for the animat's operations cannot be stored in DNA or RNA or any other methods that are discovered to be used naturally by life on Earth. The corollary: If the information necessary to execute the animat's operations can be stored in DNA or RNA, then the animat is really biological and is not an animat.

So A + B + C = a self-replicating device capable of living in our ecosphere, powered by fuel available in our ecosphere = Animat.

Assembler: A general-purpose device for molecular manufacturing capable of guiding chemical reactions by positioning molecules. A molecular machine that can be programmed to build virtually any molecular structure or device from simpler chemical building blocks. Analogous to a computer-driven machine shop.

Atomic Force Microscope (AFM) An instrument able to image surfaces to molecular accuracy by mechanically probing their surface contours. A kind of proximal probe. .... A device in which the deflection of a sharp stylus mounted on a soft spring is monitored as the stylus is moved across a surface. If the deflection is kept constant by moving the surface up and down by measured increments, the result (under favorable conditions) is an atomic-resolution topographic map of the surface. Also termed a scanning force microscope.

Atomic Layer Deposition(ALD) A self-limiting, sequential surface chemistry that deposits conformal thin-films of materials onto substrates of varying compositions. ALD is similar in chemistry to chemical vapor deposition (CVD), except that the ALD reaction breaks the CVD reaction into two half-reactions, keeping the precursor materials separate during the reaction. ALD film growth is self-limited and based on surface reactions, which makes achieving atomic scale deposition control possible. By keeping the precursors separate throughout the coating process, atomic layer control of film grown can be obtained as fine as ~ 0.1 angstroms per monolayer.

Atomic Manipulation: Manipulating atoms, typically with the tip of an STM.

Atomistic Simultations: Atomic motion computer simulations of macromolecular systems are increasingly becoming an essential part of materials science and nanotechnology. Recent advances in supercomputer simulation techniques provide the necessary tools for performing computations on nanoscale objects containing as many as 300,000 atoms and on materials simulated with 1,000,000 atoms. This new capability will allow computer simulation of mechanical devices or molecular machines using nanometer size components.

Ballistic Magnetoresistance: (BMR) is yet another way in which spin orientation, encoding information on a storage medium such as a hard drive, can modify electrical resistance in a nearby circuit, thereby accomplishing the sensing of that orientation.

Bio-Assemblies or Biomolecular Assemblies: containing several protein units, DNA loops, lipids, various ligands, etc.

Biovorous: From "biovore;" an organism capable of converting biological material into energy for sustenance.

BiochaUvinism: The prejudice that biological systems have an intrinsic superiority that will always give them a monopoly on self-reproduction and intelligence.

BioMEMS — **MEMS** used in medicine, that use microchips.

BioNEMS — biofunctionalized nanoelectromechanical systems.

Biomimetic: Imitating, copying, or learning from nature. Nanotechnology already exists in nature; thus, nanoscientists have a wide variety of components and tricks already available.

Biomimetics: study of the structure and function of biological substances to make artificial products that mimic the natural ones.

Biomimetic Chemistry: Knowledge of biochemistry, analytical chemistry, polymer science, and biomimetic chemistry is linked and applied to research in designing new molecules, molecular assemblies, and macromolecules having biomimetic functions. These new bio-related materials of high performance, including, for example, enzyme models, synthetic cell membranes, and biodegradable polymers, are prepared, tested, and constantly improved in this division for industrial scale production.

Biomimetic Materials: Materials that imitate, copy, or learn from nature.

Biopolymeroptoelectromechanical Systems [BioPOEMS]: combining optics and microelectromechanical systems, and used in biological applications.

Biostasis: A condition in which an organism's cell and tissue structure are preserved, allowing later restoration by cell repair machines. Applicable to cryonics. See also "ischemic coma," "ametabolic coma," "biostatic coma," and "in suspension."

Blue Goo - opposite of **Grey goo**. Benificial tech, or "police" nanobots.

Bogosity Filter: A mechanism for detecting bogus ideas and propositions.

Born-Oppenheimer Approximation: permits the use of classical mechanics in modeling and thinking about molecular and atomic motions. Needless to say, this greatly simplifies the conceptual framework required for thinking about molecular machines. Once used as an argument on why MNT could not work. Since refuted:

Bose-Einstein Condensates [BEC's]: "...aren't like the solids, liquids and gases that we learned about in school. They are not vaporous, not hard, not fluid. Indeed, there are no ordinary words to describe them because they come from another world — the world of quantum mechanics."

Bottom Up: Building larger objects from smaller building blocks. Nanotechnology seeks to use atoms and molecules as those building blocks. The advantage of bottom-up design is that the covalent bonds holding together a single molecule are far stronger than the weak. Mostly done by chemists, attempting to create structure by connecting molecules.

Brownian Assembly: Brownian motion in a fluid brings molecules together in various position and orientations. If molecules have suitable complementary surfaces, they can bind, assembling to form a specific structure. Brownian assembly is a less paradoxical name for self-assembly (how can a structure assemble itself, or do anything, when it does not yet exist?).

Brownian Motion: Motion of a particle in a fluid owing to thermal agitation, observed in 1827 by Robert Brown. (Originally thought to be caused by vital force, Brownian motion in fact plays a vital role in the assembly and activity of the molecular structures of life).

Bulk Technology: Technology in which atoms and molecular are manipulated in bulk, rather than individually.

Buckminsterfullerene: See Fullerenes. A broad term covering the variety of buckyballs and carbon nanotubes that exist. Named after the architect Buckminster Fuller, who is famous for the geodesic dome, which buckyballs resemble.

Bucky Balls [AKA: C60 molecules & buckminsterfullerene] - molecules made up of 60 carbon atoms arranged in a series of interlocking hexagonal shapes, forming a structure similar to a soccer ball.

Bush Robot: A concept for robots of ultimate dexterity, they utilize fractal branching to create ever-shrinking "branches," eventually ending in nanoscale "fingers." Developed by Hans Moravec.

Carbon Nanotubes: see Nanotubes

Cellular Automata: an array of identically programmed automata, or "cells," which interact with one another.

Cell pharmacology: Delivery of drugs by medical nanomachines to exact locations in the body.

Cell Repair Machine: Molecular and nanoscale machines with sensors, nanocomputers and tools, programmed to detect and repair damage to cells and tissues, which could even report back to and receive instructions from a human doctor if needed.

Chemical Vapour Deposition (CVD): a technique used to deposit coatings, where chemicals are first vaporized, and then applied using an inert carrier gas such as nitrogen.

Cobots: Collaborative robots designed to work alongside human operators. Prototype cobots are being used on automobile assembly lines to help guide heavy components like seats and dashboards into cars so they don't damage auto body parts as workers install them

Cognotechnology: Convergence of nanotech, biotech and IT, for remote brain sensing and mind control. [Nanodot]

Computational Nanotechnology: permits the modeling and simulation of complex nanometer-scale structures. The predictive and analytical power of computation is critical to success in nanotechnology: nature required several hundred million years to evolve a functional "wet" nanotechnology; the insight provided by computation should allow us to reduce the development time of a working "dry" nanotechnology to a few decades, and it will have a major impact on the "wet" side as well.

Computronium: A highly (or optimally) efficient matrix for computation, such as dense lattices of nanocomputers or quantum dot cellular automata.

Contelligence: (Consciousness + intelligence) The combination of awareness and computational power required in an Artificially Intelligent network before we could, without loss of anything essential, upload ourselves into them.

Convergent Assembly: "...rapidly make products whose size is measured in meters starting from building blocks whose size is measured in nanometers. It is based on the idea that smaller parts can be assembled into larger parts, larger parts can be assembled into still larger parts, and so forth. This process can be systematically repeated in a hierarchical fashion, creating an architecture able to span the size range from the molecular to the macroscopic."

Dendrimers: From the Greek word dendra - tree, a dendrimer is polymer that branches. "...a tiny molecular structure that interacts with cells, enabling scientists to probe, diagnose, cure or manipulate them on a nanoscale." Invented by Professor Donald Tomalia from Central Michigan University. See this article for a great explanation Dendrimers: Branching out into new realms of molecular architecture.

Design Ahead: The use of known principles of science and engineering to design systems that can only be built with tools not yet available; this permits faster exploitation of the abilities of new tools.

Design Diversity: A form of redundancy in which components of different design serve the same purpose; this can enable systems to function properly despite design flaws.

Diamondoid: Stuctures that resemble diamond in a broad sense, strong stiff structures containing dense, three dimensional networks of covalent bonds, formed chiefly from first and second row atoms with a valence of three or more. Many of the most useful diamondoid structures will in fact be rich in tetrahedrally coordinated carbon. Materials with superior strength to weight ratio, as much as 100 to 250 times as strong as Titanium, and much lighter. Possibly used to build stronger lighter rockets and space components, or a variety of other earth-bound articles for which weight and strength are a consideration.

Dip Pen Nanolithography: An AFM-based soft-lithography technique.

Directed-Assembler: A specific type of assembler that makes use of directed-assembly, such that the assembly process requires external energy or information input

Disassembler: An instrument able to take apart structures a few atoms at a time, recording structural information at each step. This could be used for uploading, copying objects (with an assembler), a dissolving agent or a weapon.

Disasterbation: Idly fantasizing about possible catastrophes (ecological collapse, full-blown totalitarianism) without considering their likelihood or considering their possible solutions and preventions.

Disruptive Technology: Technology that is significantly cheaper than current, is much higher performing, has greater functionality, and is frequently more convenient to use. Will revolutionize markets by superseding existing technology. "Paradigm shifting" is a well-worn connotation. Although the term may sound negative to some, it is in fact neutral. It is only negative when businesses who are unprepared for change fail to adapt, only to fall behind and fail. The results are not evolutionary, they are revolutionary.

Distributed Intelligence: An intelligent entity which is distributed over a large volume (or inside another system, like a computer network) with no distinct center. This is the opposite to the strategy of Concentrated intelligences. Distributed intelligences have much longer communications lags, but are more flexible in their structure and can survive damage to their parts.

DNA Chip: also: Gene Chip and DNA Microchip. A purpose built microchip used to identify mutations or alterations in a gene's DNA. See DNA Chip Technology

Dopeyballs: Superconducting Buckyballs (they) have the highest critical temperature of any known organic compound.

Dry Nanotechnology: derives from surface science and physical chemistry, focuses on fabrication of structures in carbon (e.g. fullerenes and nanotubes), silicon, and other inorganic materials. Unlike the "wet" technology, "dry" techniques admit use of metals and semiconductors.

The active conduction electrons of these materials make them too reactive to operate in a "wet" environment, but these same electrons provide the physical properties that make "dry" nanostructures promising as electronic, magnetic, and optical devices. Another objective is to develop "dry" structures that possess some of the same attributes of the self-assembly that the wet ones exhibit.

DumbSizing: apealing to the least common denominator by explaining difficult concepts in such a manner so they loose meaning. Also, talking down to someone less informed or learned.

Dyson Scenario : Life expands into the universe, which is open. As the universe cools, life stores energy to survive (do information processing). It waits until the universe is cool enough, performs some processing with part of its energy stores, then waits until the universe has cooled so much that the remaining energy can be used to do an equal amount of computation, and so on. Essentially life has to adapt as the universe grows older, changing itself to be able to survive when the stars grow cold. If the universe is open, there will be plenty of time to work in, but energy will become very scarce. Dyson has shown that a finite amount of energy is enough to guarantee infinite survival if it is spent sufficiently slowly

Dyson Sphere: A shell built around a star to collect as much energy as possible, originally proposed by Freeman Dyson (although he admits to have borrowed the concept from Olaf Stapledon's novel Star Maker (1937)). In the original proposal the shell consists of many independent solar collectors and habitats in separate orbits (also known as a Type I Dyson Sphere), but later people have discussed rigid shells consisting of only one piece (called a Type II Dyson Sphere). The latter construction is unfortunately both unstable (since it will experience no net attraction of the star), requires super-strong materials and have no internal gravity. The Dyson Sphere is a classic example of mega-technology and common in Science Fiction.

Ecophagy: (or Global Ecophagy) Consuming the biological environment. Coined and defined by Robert A. Freitas Jr. (Research Scientist Zyvex Corp). Frequently associated with "gray goo," as ecophagy (uncontrolled self-replication) is its main prupose. See "Some Limits to Global Ecophagy by Biovorous Nanoreplicators, with Public Policy Recommendations" where Dr. Freitas said "Perhaps the earliest-recognized and best-known danger of molecular nanotechnology is the risk that self-replicating nanorobots capable of functioning autonomously in the natural environment could quickly convert that natural environment (e.g., "biomass") into replicas of themselves (e.g., "nanomass") on a global basis, a scenario usually referred to as the 'gray goo problem' but perhaps more properly termed 'global ecophagy.'"

Ecosystem protector: A nanomachine for mechanically removing selected imported species from an ecosystem to protect native species.

Electrical Bistability: a phenomenon in which an object exhibits two states of different conductivity at the same applied voltage.

Emergence: a complex whole created by simple parts, as in the brain where billions of neurons work individually, but collectively make up our consciousness and give us the ability to think, rationalize, and create.

EI - Emergent Intelligence: An intelligent system that gradually emerges from simpler systems, instead of being designed top down.

Emulation: An absolutely precise simulation of something, so exact that it is equivalent to the original (for example, many computers emulate obsolete computers to run their programs). The Star Trek replicator is an example.

Enabling science and technologies: Areas of research relevant to a particular goal, such as nanotechnology. Also, technology that "enables" other technology to advance, such as the transistor enabled the computer chip revolution, as did photolithography.

Entanglement: From quantum mechanics, entanglement is a relationship between two objects in which they both exhibit superposition but once the state of one object is measured, the state of the other is also known.

Entropy: A measure of the disorder of a closed system. The second law of thermodynamics states that the entropy (and disorder) increases as time moves forward.

Evolution: A process in which a population of self-replicating entities undergoes variation, with successful variants spreading and becoming the basis for further variation.

Exploratory Engineering: Design and analysis of systems that are theoretically possible but cannot be built yet, owing to limitations in available tools.

Exponential Assembly: a manufacturing architecture starting with a single tiny robotic arm on a surface. This first robotic arm makes a second robotic arm on a facing surface by picking up miniature parts ó carefully laid out in advance in exactly the right locations so the tiny robotic arm can find them ó and assembling them. The two robotic arms then make two more robotic arms, one on each of the two facing surfaces. These four robotic arms, two on each surface, then make four more robotic arms. This process continues with the number of robotic arms steadily increasing in the pattern 1, 2, 4, 8, 16, 32, 64, etc. until some manufacturing limit is reached (both surfaces are completely covered with tiny robotic arms, for example). This is an exponential growth rate, hence the name exponential assembly. See Exponential Assembly

Exponential Growth: inaccurately referred to as "self-replication," exponential growth refers to the process of growth or replication involving doubling within a given period.

Femtometer: [abbr: fm] a unit suitable to express the size of atomic nuclei. One quadrillionth (10 to minus 15) of a meter.

Femtosecond: is one quadrillionth of a second, and is to a second what a second is to 32,700,000 years. At 186,000 miles per second, in one femtosecond light travels only far enough to traverse about 1,000 silicon atoms. When used to time a laser pulse, it allows for ultra-precise micromachining, with virtually no damage to surrounding material.

Femtotechnology: the art of manipulating materials on the scale of elementary particles (leptons, hadrons, and quarks). The next step smaller after picotechnology, which is the next step smaller after nanotechnology.

Fluidic Self Assembly: A novel technique for accurately assembling large numbers of very small devices. The small size, planarity, and accuracy of the assembly also result in very low parasitic interconnects, comparable to on die traces. This massively parallel assembly process combines the capability and flexibility of assembly with the cost effectiveness of integration.

Invented by Mr. Mark Hadley and was part of his Ph.D. dissertation while he was studying at University of California, Berkley. The FSA process became the foundation for the origins of a new company named Alien Technology Corporation. In the FSA process, specifically shaped semiconductor devices ranging in size from 10 microns to several hundred microns are suspended in liquid and flowed over a surface which has correspondingly shaped "holes" or receptors on it and into which the devices settle. The shape of the devices and of the holes is designed so that the devices fall easily into place and are selfaligning. Alien has successfully demonstrated the assembly of tens of thousands of devices in a single process step.

Foglet: A mesoscale machine. A discreet component of utility fog.

Fractal: A mathematical construct that has a fractional dimension. See Fractal eXtreme Gallery & Fractal Domains for examples [images] and software to create your own.

Fractal Mechatronic Universal Assembler: (or Fractal Assembler) is a machine that is capable of assembling any chemical from a generic descriptions of the properties required of the chemical. The machine comprises of test tube arrays and software linked to robotic cubes and sensor arrays to implement automated mixing and testing to conduct materials research activity. [FR] See Fractal Mechatronic Universal Assembler

Fractal Robots: AKA: Fractal Shape Shifting Robots and Programmable "Digital Matter", are programmable machines that can do unlimited tasks in the physical world, the world of matter. Load the right software and the same "machines" can vacuum the carpet, paint your car, or construct an office building and later, wash that building's windows. This is the beginning of "Digital Matter".

Fractal Shape Shifting Robots look like "Rubic's Cubes" that can "slide" over each other on command, changing and moving in any overall shape desired for a particular task. These cubes communicate with each other and share power through simple internal induction coils (or surface contacts in some models), have batteries, a small computer and various kinds of internal magnetic and electric inductive motors (depending on size) used to move over other cubes.

When sufficiently miniaturized (below 0.1mm) and fabricated using photolithography and E-Beam methods, the machines may exceed human manual dexterity and could then be programmed to assemble complex fractal aggregates or even to maintain the photolithographic and E-Beam equipment itself! The ultimate goal is self sustaining systems and "self-assembly" features that can drop cost dramatically and enable successive generations of robots exhibiting greater utility and value, to be built along the way.

FUD: Fear, Uncertainty, Doubt.

Fullerenes: Fullerenes are a molecular form of pure carbon discovered in 1985. They are cage-like structures of carbon atoms, the most abundant form produced is buckminsterfullerene (C60), with 60 carbon atoms arranged in a spherical structure. There are larger fullerenes containing from 70 to 500 carbon atoms. [Wid] See What are fullerenes?

Genegeneering: Genetic engineering.

Genetic Algorithm: Any algorithm which seeks to solve a problem by considering numerous possibilities at once, ranking them according to some standard of fitness, and then combining ("breeding") the fittest in some way. In other words, any algorithm which imitates natural selection.

GENIE: An AI combined with an assembler or other universal constructor, programmed to build anything the owner wishes. Sometimes called a Santa Machine. This assumes a very high level of AI and nanotechnology.

Golden Goo: Another member of the grey goo family of nanotechnology disaster scenarios. The idea is to use nanomachines to filter gold from seawater. If this process got out of control we would get piles of golden goo (the "Wizard's Apprentice Problem"). This scenario demonstrates the need of keeping populations of self-replicating machines under control; it is much more likely than grey goo, but also more manageable.

GNR technologies (Genetic Engineering, Nanotechnology, and Robotics)

Gray Goo or Grey Goo - destructive **nanobots** [AKA: "**gray dust**"]. opposite of Blue Goo. See Star Trek scenario. Vast legions of destructive nanites. Typically, created by accident. Left unchecked, they will basically convert everthing they contact into more of themselves, or consume and digest it for energy. Either way, its pretty much bad news. The debate rages on.

Also - Self-replicating (von Neumann) nanomachines spreading uncontrolably, building copies of themselves using all available material. This is a commonly mentioned nanotechnology disaster scenario, although it is rather unlikely due to energy constraints and elemental abundances. More probable disaster scenarios are the green goo, golden goo, red goo, khaki goo scenarios. As a protection blue goo has been proposed.

Green Goo: Nanomachines or bio-engineered organisms used for population control of humans, either by governments or eco-terrorist groups. Would most probably work by sterilizing people through otherwise harmless infections. See Nick Szabo's essay Green Goo -- Life in the Era of Humane Genocide.

Guy Fawkes Scenario: If nanotechnology becomes widely available, it might become trivial for anyone to committ acts of terrorism (such as making nanomachines build a large amount of explosives under government buildings a la Guy Fawkes). This would either force strict control over nanotechnology (hard) or a decentralized mode of organization.

Heisenberg Uncertainty Principle: A quantum-mechanical principle with the consequence that the position and momentum of an object cannot be precisely determined. The Heisenberg principle helps determine the size of electron clouds, and hence the size of atoms. [NTN] "The more precisely the POSITION is determined, the less precisely the MOMENTUM is known"

Heteronuclear: consisting of different elements.

Intelligent Agent: aka "software agent". Software that can do things without supervision, because it knows your patterns, history, preferences, likes, dislikes, and so forth. You want to take a vacation - it knows that you really enjoyed that trip to Hawaii, and that you prefer to fly at night, 1st class. It also knows that the bungalow you rented last time was marked as being 5-star, and worth a re-visit. Your IA then collates all your parameters, searches the internet for flights, car rentals, restaurant reservations, and lodgings, and schedules everything for you, with options on the side. No more travel agent - you have a software agent to handle things! Many experts agree that in near future we will each have one, and that they will greatly reduce our daily load of trivial and redundant tasks.

IA: Intelligence Amplification: Technologies seeking to increase the cognitive abilities of people.

Immune Machines: Medical nanomachines designed for internal use, especially in the bloodstream and digestive tract, able to identify and disable intruders such as bacteria and viruses.

IMP: Electronic implant, especially in the brain.

Inline Universities: (as opposed to online universities), nanocomputer implants serving to increase intelligence and education of their owners, essentially turning them into walking universities [Max M. Rasmussen]

Jupiter-Brain: A posthuman being of extremely high computational power and size. This is the archetypal concentrated intelligence. The term originated due to an idea by Keith Henson that nanomachines could be used to turn the mass of Jupiter into computers running an upgraded version of himself.

ki Goo: Military nanotechnology; see grey goo.

Knowbots: Knowledge robots, first developed Vinton G. Cref and Robert E. Kahn for National Research Initiatives. Knowbots are programmed by users to scan networks for various kinds of related information, regardless of the language or form in which it expressed. "Knowbots support parallel computations at different sites. They communicate with one another, and with various servers in the network and with users." [Scientific American, September 1991, p.74.]

Langmuir-Blodgett: The name of a nanofabrication technique used to create ultrathin films (monolayers and isolated molecular layers), the end result of which is called a "Langmuir-Blodgett film." More and more.

LCD (Liquid Crystal Display) is the predominant technology used in flat panel displays. The principle that makes the display work is this: A crystalís alignment can be altered with an electric current. If the crystal is lined up one way ñ it will allow the light waves to pass through a polarized filter, but if the electric current alters the crystalís alignment, it will guide light so that the polarized filter blocks the light. By densely packing red, blue and green light emitting crystals next to each other on a sheet (ìcalled a substrateî), one can create a full color display. The great thing about LCD is that the crystals can be packed together closely, allowing for a higher-resolution, finer-detail display. The con is that LCDs are somewhat fragile, require a lot of power and are relatively less bright.

LEDs (Light Emitting Diodes) work on a completely different concept. Traditionally LEDs are created from two semiconductors. By running current in one direction across the semiconductor the LED emits light of a particular frequency (hence a particular color) depending on the physical characteristics of the semiconductor used. The semiconductor is covered with a piece of plastic that focuses the light and increases the brightness. These semiconductors are very durable, there is no filament, they donít require much power, theyíre brighter and they last a long time. By densely packing red, blue and green LEDs next to each other on a substrate one can create a display.

The disadvantage of LEDs is that they are much larger ñ therefore the resolution is not nearly as good as LCD displays. Thatís why most LED displays are large, outdoor displays, not smaller devices, like monitors.

Limited Assembler: Assembler capable of making only certain products; faster, more efficient, and less liable to abuse than a general-purpose assembler.

Linde Scenario: A scenario for indefinite survival of intelligent life. It assumes it is possible to either create basement universes connected to the original universe with a wormhole or the existence of other cosmological domains. Intelligent life continually migrates to the new domains as the old grow too entropic to sustain life. [AS/Mitch Porter, 1997. The name refers to Linde's chaotic inflation cosmology, where new universes are continually spawned.] See The Linde scenario, v0.01

Lofstrom Loop: An beanstalk-like megaconstruction based on a stream of magnetically accelerated bars linked together. The stream is sent into space, where a station rides it using magnetic hooks, redirects it horizontally to another station, which sends it downwards to a receiving station on the ground. From this station the stream is then sent back to the launch station (a purely vertical version is called a space fountain). This structure would contain a large amount of kinetic energy but could be built gradually and would only require enough energy to compensate for losses when finished. Elevators could be run along the streams, and geostationary installations could be placed along the horizontal top.

Low-dimension Structures: quantum wells, quantum wire and quantum dots.

Matter as Software: "Autonomous, motile microdevices clearly are on the horizon. They may be regarded as the first step in the evolution of a technology for "programming" the structure and properties of material objects at the microscopic and the submicroscopic levels. As this evolution progresses, the physical and economic properties of such programmable matter are likely to become much like those of present day software."

Meat Machine: [AKA Cabinet Beast]. A box containing assemblers and raw material, within which is formed meat [or whatever else it was programmed to make].

Mechanochemistry: the direct, mechanical control of molecular structure formation and manipulation to form atomically precise products

Mechanosynthesis: (where) molecular tools with chemically specific tip structures can be used, sequentially, to modify a work piece and build a wide range of molecular structures.

Mechatronics: the study of the melding of AI and electromechanical machines to make machines that are greater than the sum of their parts.

Meme: An idea that replicates through a society as it is propagated through person-to-person interaction, both direct and indirect. Memetics is a field of study that focuses on memes' role in the evolution of a culture.

MEMS--MicroelectroMechanical Systems: generic term to describe micron scale electrical/mechanical devices.

Mesoscale: A device or structure larger than the nanoscale (10^-9 m) and smaller than the megascale; the exact size depends heavily on the context and usually ranges between very large nanodevices (10^-7 m) and the human scale (1 m).

Microencapsulation: Individually encapsulated small particles.

MIMIC: [micromoulding in capillaries] one-step rapid prototyping technique.

Molecular Assembler: Also known as an assembler, a molecular assembler is a molecular machine that can build a molecular structure from its component building blocks.

Molecular Beam Epitaxy: [MBE] Process used to make compound (multi-layer) semiconductors. Consists of depositing alternating layers of materials, layer by layer, one type after another (such as the semiconductors gallium arsenide and aluminum gallium arsenide).

Molecular Biology: [AKA: wet nano]

Molecular Integrated Microsystems (MIMS): microsystems in which functions found in biological and nanoscale systems are combined with manufacturable materials. See Molecular Integrated Microsystems

Molecular Electronics (ME) [moletronics] Any system with atomically precise electronic devices of nanometer dimensions, especially if made of discrete molecular parts rather than the continuous materials found in today's semiconductor devices. Also: Using molecule-based materials for electronics, sensing, and optoelectronics .... ME is the set of electronic behaviors in molecule-containing structures that are dependent upon the characteristic molecular organization of space ME behavior is fixed at the scale of the individual molecule, which is effectively the nanoscale.

Molecular Manipulator: A device combining a proximal probe mechanism for atomically precise positioning with a molecule binding site on the tip; can serve as the basis for building complex structures by positional synthesis.

Molecular Manufacturing: Manufacturing using molecular machinery, giving molecule-by-molecule control of products and by-products via positional chemical synthesis.

Molecular Medicine: Studying molecules as they relate to health and disease, and manipulating those molecules to improve the diagnosis, prevention, and treatment of disease. [see Medscape Molecular Medicine for news]

Molecular Nanogenerator: see Molecular Nanogenerator developed that can target cancer cells and destroy them.

Molecular Nanotechnology (MNT): Thorough, inexpensive control of the structure of matter based on molecule-by-molecule control of products and byproducts; the products and processes of molecular manufacturing, including molecular machinery.

Molecular Recognition: A chemical term referring to processes in which molecules adhere in a highly specific way, forming a larger structure; an enabling technology for nanotechnology. [FS]

Molecular Systems Engineering: Design, analysis, and construction of systems of molecular parts working together to carry out a useful purpose.

Molecular Wire: A molecular wire - the simplest electronic component - is a quasi-one-dimensional molecule that can transport charge carriers (electrons or holes) between its ends. [Michael D Ward]

MOLMAC: Molecular machine [Kilian, Gryphon]

Monomer: The units from which a polymer is constructed.

Monomolecular Computing: the implantation inside a single molecule of ALL the functional groups or circuits to realize a calculation, without any help from external artifices such as re-configuration, calculation sharing between the user and the machine, or selection of the operational devices.

Moore's Law -- Coined in 1965 by Gordon Moore, future chairman and chief executive of Intel, it stated at the time that the of number transistors packed into an integrated circuit had doubled every year since the technology's inception four years earlier. In 1975 he revised this to every two years, and most people quote 18 months. The trend cannot continue indefinitely with current lithographic techniques, and a limit is seen in ten to fifteen years. However, the baton could be passed to nanoelectronics, to continue the trend (though the smoothness of the curve will very likely be disrupted if a completely new technology is introduced).

Nanarchist: Someone who circumvents government control to use nanotechnology, or someone who advocates this.

Nanarchy: The use of automatic law-enforcement by nanomachines or robots, without any human control

Nanoarray: an ultra-sensitve, ultra-miniaturized array for biomolecular analysis. BioForce Nanosciences' Nanoarrays utilize approximately 1/10,000th of the surface area occupied by a conventional microarray, and over 1,500 nanoarray spots can be placed in the area occupied by a single microarray domain.

Nanoassembler: the Holy Grail of nanotechnology; once a perfected nanoassembler is availble, building anything becomes possible, with physics and the imagination the only limitation (of course each item would have to be designed first, which is another small hurdle).

Nanobalance: Simply put, a nanoscale balance for determining mass, small enough to weigh viruses and other sub-micron scale particles. "*A mass attached at the end of a nanotube shifts its resonance frequency. If the nanotube is calibrated (i.e., its spring constant known), it is possible to measure the mass of the attached particle.*" *A nanobalance "could be useful for determining the mass of other objects on the femtogram to picogram size range."*

Nanobarcode: SurroMed's Nanobarcode™ technology uses cylindrically-shaped colloidal metal nanoparticles, in which the metal composition can be alternated along the length and the size of each metal segment can be controlled. Intrinsic differences in reflectivity between the metal segments allow individual particles to be identified by conventional optical microscopy

Nanobeads: Polymer beads with diameters of between 0.1 to 10 micrometers. Also called nanodots, nanocrystals and quantum beads. Impregnating fluorescent crystal chips into these beads allows simultaneous measurement of thousands of biological interactions, a stepping stone for breakthroughs in the diagnosis and treatment of disease. ... with the potential to accelerate drug discovery and clinical diagnostics."

Nanobialys -- Ultra-miniature bialy-shaped particles developed by Washington University as delivery agents for drugs and imaging agents directly to the sites of tumors and plaques.

Nanobiotechnology: applying the tools and processes of MNT to build devices for studying biosystems, in order to learn from biology how to create better nanoscale devices. Should hasten the creation of useful micro devices that mimic living biological systems.

Nanobot: See Nanite

Nanobubbles: tiny air bubbles on colloid surfaces. Thought to reduce drag, such as would be of benefit to swimmers wearing a suit coverd in them.

Nanochips: we are approaching the limits of standard microchip technology; thus, the "nanochip" -- a next-smaller microchip. [ed] They are also a next-gen device for mass storage, of significantly higher density, with greater speed, and much lower cost.

Nanocomputer: A computer made from components (mechanical, electronic, or otherwise) built at the nanometer scale. These computers could be many orders-of-magnititude faster than today's, which enables software to take proportional leaps.

Nanochondria: Nanomachines existing inside living cells, participating in their biochemistry (like mitochondria) and/or assembling various structures. See also nanosome. [Ken Clements 1996]

Nanocones: Nonplanar graphitic structures. Carbon-based structures with five-fold symmetry that form due to disclination defects in two-dimensional graphene sheets. They have been observed as nanotube caps and as freestanding structures. [North Carolina State University]

Nanocontainers: "Micellar nanocontainers" or "Micelles," these are nanoscale polymeric containers that could be used to selectively deliver hydrophobic drugs to specific sites within individual cells.

Nanocrystals: also known as nanoscale semiconductor crystals. "Nanocrystals are aggregates of anywhere from a few hundred to tens of thousands of atoms that combine into a crystalline form of matter known as a "cluster." Typically around ten nanometers in diameter, nanocrystals are larger than molecules but smaller than bulk solids and therefore frequently exhibit physical and chemical properties somewhere in between. Given that a nanocrystal is virtually all surface and no interior, its properties can vary considerably as the crystal grows in size."

Nano Cubic Technology: an ultra-thin layer coating that results in higher resolution for recording digital data, ultra-low noise and high signal-to-noise ratios that are ideal for magneto-resistive (MR) heads. It is capable of catapulting data cartridge and digital videotape to one-terabyte native (uncompressed) capacities and floppy disk capacities to three gigabytes. To help visualize the potential, 1TB can store up to 200 two-hour movies. [Fuji Photo Film U.S.A., Inc.]

Nanodefenses: any of the "good" goo's, such a Blue Goo. Protectors against Grey Goo, destructive nanoswarms, and the like.

Nanodisaster: See the various 'goo' scenerios that have potentially negative outcomes.

NEMS - nanoelectromechanical systems: A generic term to describe nano scale electrical/mechanical devices.

Nanoelectronics: Electronics on a nanometer scale, whether made by current techniques or nanotechnology; includes both molecular electronics and nanoscale devices resembling today's semiconductor devices.

Nanofabrication: construction of items using assemblers and stock molecules. See Nanofacture. AKA : nanoscale engineering.

Nanofacture: The fabrication of goods using nanotechnology [Geoff Dale 1995]. See Nanofabrication

Nanofilters: One opportunity for nanoscale filters is for the separation of molecules, such as proteins or DNA, for research in genomics can be used, as "*masks to prevent exposure to biological pathogens such as viruses that can be as small as 30 nanometers in diameter.*" Another use is in water filtration.

Nanofluidics: controlling nano-scale amounts of fluids

Nanogate: A device that precisely meters the flow of tiny amounts of fluid. Precise control of the flow restriction is accomplished by deflecting a highly polished cantilevered plate. The opening is adjustable on a sub-nanometer scale, limited by the roughness of the polished plates. Thus, the Nanogate is an Ultra Surface Finish Effect Mechanism (USFEM). The Nanogate can be fabricated on a macro-, meso- or micro- (MEMs) scale. [James R. White]

Nanoguitar: "Made for fun to illustrate the technology -- the world's smallest guitar is 10 micrometers long -- about the size of a single cell -- with six strings each about 50 nanometers, or 100 atoms, wide. Just one of several structures that Cornell researchers believe are the world's smallest silicon mechanical devices. Researchers made these devices at the Cornell Nanofabrication Facility, bringing microelectromechanical devices, or MEMS, to a new, even smaller scale -- the nano-sized world.

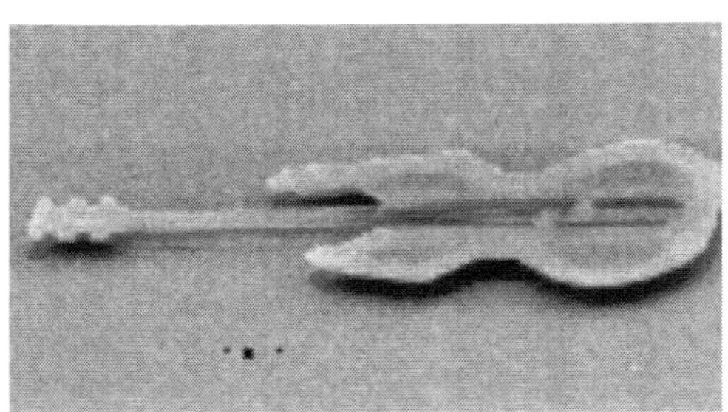

Smallest guitar, about the size of a human blood cell, carved out of crystalline silicon, illustrates new technology for nanosized electromechanical devices.

Nanogypsy: someone who travels form place to place, spreading the "nano" word. Usually a person who takes the most optimistic viewpoint, and is enthusitic.

Nanohorns: One of the SWNT (single walled carbon nanotube) types, with an irregular horn-like shape, which may be a critical component of a new generation of fuel cells. "*The main characteristic of the carbon nanohorns is that when many of the nanohorns group together an aggregate (a secondary particle) of about 100 nanometers is created. The advantage being, that when used as an electrode for a fuel cell, not only is the surface area extremely large, but also, it is easy for the gas and liquid to permeate to the inside. In addition, compared with normal nanotubes, because the nanohorns are easily prepared with high purity it is expected to become a low-cost raw material.*"

Nanoimprinting: Sometimes called soft lithography. A technique that is very simple in concept, and totally analogous to traditional mould- or form-based printing technology, but that uses moulds (masters) with nanoscale features. As with the printing press, the potential for mass production is clear. There are two forms of nanoimprinting, one that uses pressure to make indentations in the form of the mould on a surface, the other, more akin to the printing press, that relies on the application of "ink" applied to the mould to stamp a pattern on a surface. Other techniques such as etching may then follow.

Nanoimprint Machine: a form of soft lithography

Nanohacking: describes what MNT is all about -- "hacking" at the molecular level.

Nanoindentation: Nanoindentation is similar to conventional hardness testing performed on a much smaller scale. The force required to press a sharp diamond indenter into a material is measured as a function of indentation depth. As depth resolution is on the scale of nanometers (hence the name of the instrument), it is possible to conduct indentation experiments even on thin films. Two quantities which can be readily extracted from nanoindentation experiments are the material's modulus, or stiffness, and its hardness, which can be correlated to yield strength. Investegators have also used nanoindentation to study creep, plastic flow, and fracture of materials.

Nanolithography: Writing on the nanoscale. From the Greek words Nanos - Dwarf, Lithos - rock, and grapho - to write, this word literally means "small writing on rocks."

Nanomachine: An artificial molecular machine of the sort made by molecular manufacturing.

Nanomachining: like traditional machining, where portions of the structure are removed or modified, nanomachining involves changing the structure of nano-scale materials or molecules.

NanoManipulator: uses virtual reality (VR) goggles and a force feedback probe as an interface to a scanning probe microscope, providing researchers with a new way to interact with the atomic world. Researchers can travel over genes, tickle viruses, push bacteria around, and tap on molecules - the nanoManipulator simplifies the process and allows researchers to play with their atoms.

Nanomanipulation: The process of manipulating items at an atomic or molecular scale in order to produce precise structures.

Nanomanufacturing: Same as molecular manufacturing.

Nanomaterials: can be subdivided into nanoparticles, nanofilms and nanocomposites. The focus of nanomaterials is a bottom up approach to structures and functional effects whereby the building blocks of materials are designed and assembled in controlled ways.

Nanomesh and Nanofibres: (or "Nanofibers") This term covers CNT's (see above), and as described here, the other "nanoscale fibers" referred to as "polymeric" (made from polymers). Currently used in air and liquid filtration applications. Using a process called "electrospinning" - or e-spin - a polymer "mesh" is formed into a nanofiber membrane, hense "nanomesh", with 150 - 200 nm diameters. Some have been made since 1970, but were not called "nano" until recently. One potential use is "*to prevent body tissues from sticking together as they heal. It also breaks down in the body over time like biodegradable sutures.*" , which makes it a surgical material for the 21st Century. Other uses include biomedical devices, filtration systems, and dust collecting systems.

Nano-Optics: Interaction of light and matter on the nanoscale.

Nanopens & Nanopencils: (AKA: Atomic Pencil) "Analogous to using a quill pen but on a billionth the scale", and may transform dip-pen nanolithography. Allows for drawing electronic circuits a thousand times smaller than current ones. The "pen" is an atomic force microscope (AFM). See Nanopipettes and Nanoplotter for further details.

NanoPGM - nanometer-scale patterned granular motion: The goal of NanoPGM is to generate millions of ìnanofingers,î finger-like structures each only a few nanometers long, that might someday perform precise, massively parallel manipulation of molecules and directed assembly of other nanometer-scale objects. This ability answers one of the biggest technical challenges facing builders of nanocomputers: how to arrange as many as a trillion molecular computing components in an area only a few millimeters square.

Nanopharmaceuticals: nanoscale particles used to modulate drug transport for drug uptake and delivery applications.

Nanophase Carbon Materials (carbon nanotubes, nanodiamond, nanocomposite]--A form of matter in which small clusters of atoms form the building blocks of a larger structure. These structures differ from those of naturally occurring crystals, in which individual atoms arrange themselves into a lattice.

Nanopipettes: "Cantilevered/Straight Nanopipettes can be used as nanopens for controlled chemical delivery or removal from regions as small as 100 nanometers. They can also be used as vessels for containing molecules whose optical properties change in response to their chemical environment." Other uses include "controlled chemical etching with the precision of atomic force microscopy; chemical imaging of surfaces; delivering femtosecond laser pulses; and performing NSOM/SNOM imaging using a UV excimer laser."

Nanoplotter: A multi-tip nanopen. "A device that can draw patterns of tiny lines just 30 molecules thick and a single molecule high. ... produces eight identical patterns at once and extends ... dip-pen nanolithography towards mass producing nanoscale devices and circuits by converting what was a serial process to a parallel one. May be use to "... miniaturize electronic circuits, pattern precise arrays of organic and biomolecules such as DNA and put thousands of different medical sensors on an area much tinier than the head of a pin."

Nanopores: Involves squeezing a DNA sequence between two oppositely charged fluid reservoirs, separated by an extremely small channel. Essentially itty bitty tiny holes. Nanoscopic pores found in purpose-built filters, sensors, or diffraction gratings to make them function better. See Influencing structure in the heart of nanoland. As activated carbon, they may also be used as an alternative fuel storage medium, due to their massive internal surface area. "Scientists believe nanopores, tiny holes that allow DNA to pass through one strand at a time, will make DNA sequencing more efficient." See Understanding Nanodevices -- Nanopores. In biology, they are "complex protein assemblies that span cell membranes and allow ionic transport across the otherwise impermeable lipid bilayer. Nanopores are important because while some pores help maintain cell homeostasis, others disrupt cell function." See Towards Fabrication of Solid-State Mimics of Biological Nanopores. "A nanopore can be a protein channel in a lipid bilayer or an extremely small isolated 'hole' in a thin, solid-state membrane" such that "DNA and RNA, can be registered and characterized singly ...

Nanoprobe: Nanoscale machines used to diagnose, image, report on, and treat disease within the body. See "Cell Repair Machine", "Nanites", "Nanobots", and "Nanomachine". Also: tips for scanning probe microscopes.

Nanoreplicators: A set of nanomachines capable of exponential replication.

Nanorods: or Carbon Nanorods. Formed from multi-wall carbon nanotubes. Another nanoscale material with unique and promising physical properties, such that may yield improvements in high-density data storage, and allow for cheaper flexible solar cells.

Nanoropes: nanotubes connected and strung together.

Nanoscale: 1 - 100 nanometer range.

Nanoscopic Scale same as nanoscale.

Nanosensors: nanoscale sensors.

Nanoshells: Nanoscale metal spheres, which can absorb or scatter light at virtually any wavelength. "The nanoshells act as an amazingly versatile optical component on the nanometer scale: they may provide a whole new approach to optical materials and components''.

Nanosources: sources that emit light from nanometre-scale volumes.

Nanosome: Nanodevices existing symbiotically inside biological cells, doing mechanosynthesis and disassembly for it and replicating with the cell. Similar to nanochondria.

Nanosprings: A nanowire wrapped into a helix. Speculation is that they "may someday make highly sensitive magnetic field detectors, perhaps finding application in hard drive read heads. Alternatively, nanosprings could serve as positioners, or even as tiny conventional springs, for nanomachines of the future."

Nanosurgery: A generic term including molecular repair and cell surgery.

Nanoswarm: UFog and Goo

Nanotechism: the religion of nanotech, as opposed to the science of nanotech

Nanotechnology: a manufacturing technology able to inexpensively fabricate most structures consistent with natural law, and to do so with molecular precision. [FS]

Nanoterrorism: using MNT derived nanites to do damage to people or places.

Nano-test-tubes: CNT's opened and filled with materials, and used to carry out chemical reactions.

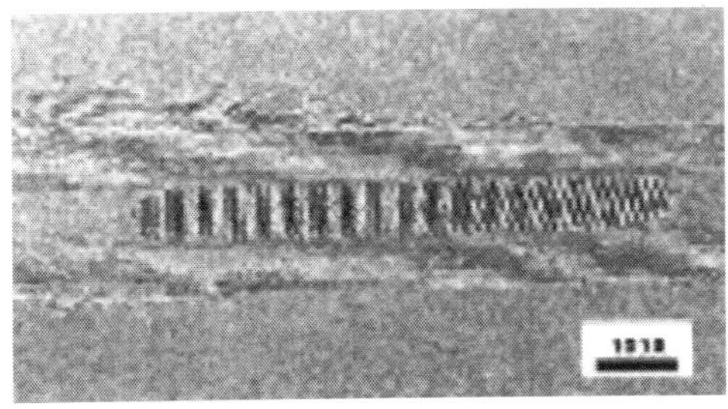

A High Resolution Transition Electron Micrograph (TEM) of Samarium Oxide Inside a Multi-Walled Carbon Nanotube.

Nanotribology: the methodical study of friction, lubrication, and wear at the nanoscale.

Nanotube: A one dimensional fullerene (a convex cage of atoms with only hexagonal and/or pentagonal faces) with a cylindrical shape. Carbon nanotubes discovered in 1991 by Sumio Iijima resemble rolled up graphite, although they can not really be made that way. Depending on the direction that the tubes appear to have been rolled (quantified by the 'chiral vector'), they are known to act as conductors or semiconductors. Nanotubes are a proving to be useful as molecular components for nanotechnology.

Strictly speaking, any tube with nanoscale dimensions, but generally used to refer to carbon nanotubes (a commonly mentioned non-carbon variety is made of boron nitride), which are sheets of graphite rolled up to make a tube. The dimensions are variable (down to 0.4 nm in diameter) and you can also get nanotubes within nanotubes, leading to a distinction between multi-walled and single-walled nanotubes. Apart from remarkable tensile strength, nanotubes exhibit varying electrical properties (depending on the way the graphite structure spirals around the tube, and other factors), and can be insulating, semiconducting or conducting (metallic).

"Maybe the most significant spin-off product of fullerene research....are nanotubes based on carbon or other elements. These systems consist of graphitic sheets seamlessly wrapped to cylinders. With only a few nanometers in diameter, yet (presently) up to a millimeter long, the length-to-width aspect ratio is extremely high. A truly molecular nature is unprecedented for macroscopic devices of this size. Accordingly, the number of both specialized and large-scale applications is growing constantly."

Nanowires: "Semiconductor nanowires are one-dimensional structures, with unique electrical and optical properties, that are used as building blocks in nanoscale devices." See Nanowires within nanowires and Learning how to Fabricate Nanowire. "Striped or 'superlatticed' nanowires can function as transistors, LEDs (light-emitting diodes) and other optoelectronic devices, biochemical sensors, heat-pumping thermoelectric devices, or all of the above, along the same length of wire."

Nanny: A cell-repair nanite

NEMS - Nanoelectromechanical systems: Nanoscale MEMS.

nm: Abbreviation for Nanometer.

NRAM™ - Nanotube-based/Nonvolatile RAM, developed by Nantero, using proprietary concepts and methods derived from leading-edge research in nanotechnology.

Nanowetting: how wetting behavior depends on nanoscale topography on a substrate. [BNL]

NBIC: Nanotechnology, Biotechnology, Information Technology and Cognitive Science.

NE^{3}LS: Nanotechnology's Ethical, Environmental, Economic, Legal, and Social Implications.

OLED or **Organic LED** is not made of semiconductors. It's made from carbon-based molecules. That is the key science factor that leads to potentially eliminating LEDs' biggest drawback - size. The carbon-based molecules are much smaller. And according to a paper written by Dr. Uwe Hoffmann,

Dr. Jutta Trube and Andreas Kl-ppel, entitled OLED - A bright new idea for flat panel displays "OLED is brighter, thinner, lighter, and faster than the normal liquid crystal (LCD) display in use today. They also need less power to run, offer higher contrast, look just as bright from all viewing angles and are - potentially - a lot cheaper to produce than LCD screens."

OMEGA POINT: Also called the Quantum Omega Point Theory. A possible future state when intelligence controls the Universe totally, and the amount of information processed and stored goes asymptotically towards infinity.

Orbital Tower: also known as a "space tether", "beanstalk" or "heavenly funicular". A cable in synchronous orbit, with one end anchored to the surface of the Earth, often with a small asteroid at the outer end to provide some extra tension and stability. Picture also a "space elevator".

In theory, constructed of a diamondoid material, approximately 22,000 miles long, with one end in a stable orbit, and the other somewhere [probably] around the equator. Used frequently in science-fiction yarns, and may become a reality with the advent of mature MNT. Such an elevator would move freight and passengers into orbit at a cost per pound orders of magnitude less than current launches, with passenger safety comparable to train, plane, or subway trips. Becomes possible when we can mass-produce nanotubes, and make their length to fit.

Paradigm Shift: When one conceptual world-view is replaced by another, or, a change of patterns on a massive scale. When Copernicus showed how the Earth rotates around the Sun, and not vice versa, that created a paradigm shift [it forced a new way of thinking about our place in the Universe]. And when quantum physics and general relativity displaced Newtonian mechanics, that created another shift. Applied to an enabling technology such as molecular manufacturing, it suggests that there will be many shifts occurring, soon, and with wide-ranging and often disruptive consequences.

Pervasive Computing: when computers (and sensors and actuators) become virtually invisible, and are used in almost every aspect of human commerce, interaction, and life. It will allow you full control over data and information, enabling you to send, receive, manage, and update your data from anywhere at any time. It will also allow you full control over your environment, in so far as you will be able to speak or gesture commands, effecting changes to things around you. Applications include: environmental monitoring - when you enter a room, they sense your presence and adjust temperature and humidity to your personal preferences; building security - to sense chemical weapons and perform face recognition; information transfer - allowing you to send and receive calls, data, and images from anywhere to anywhere, without the need of bulky equipment. Also called **"Ubiquitous Computing", "Intelligent Telesensing", "Proactive Computing", "Distributed Information Management Systems", "The Evernet",** and "Calm Technology". "...it will look like nothing to the naked eye. ...beneath the surface, tiny computing networks will be doing exactly what we want them to do - working behind the scenes to help us see clearer, travel safer, and place more knowledge, rather than frustration, into our heads."

Pico Technology: (trillionth of a meter) -- the next step smaller, after Nanotechnology. The art of manipulating materials on a quantum scale.

Pink Goo: (humorous) Humans (in analogy with grey goo). "Pink Goo to refer to Old Testament apes who see their purpose as being fruitful and multiplying, filling up of the cosmos with lots more such apes, unmodified."

POSS Nanotechnology: short for Polyhedral Oligomeric Silsesquioxanes Nanotechnology. POSS nanomaterials are attractive for missile and satellite launch rocket applications because they offer effective protection from collisions with space debris and the extreme thermal environments of deep space and atmospheric re-entry. Another application of POSS nanotechnology under development is a new high-temperature lubricant. This new nanolubricant is effective at temperatures up to 500ƒF, which is 100 F greater than conventional lubricants.

Polysilicon: short for Polycrystalline Silicon, used in the manufacture of computer chips.

Posthuman: Persons of unprecedented physical, intellectual, and psychological capacity, self-programming, self-constituting, potentially immortal, unlimited individuals.

Positional Controlled Chemical Synthesis or Positional Synthesis: Control of chemical reactions by precisely positioning the reactive molecules, the basic principle of assemblers.

Positional Assembly: Constructing materials an atom or molecule at a time

Post Monetary Economy: After the advent of mature Nanotechnology, it is likely that our economic reality will change, possibly to the extent of eliminating currency as we know it today.

Protein Design, Protein Engineering: The design and construction of new proteins; an enabling technology for nanotechnology.

Protein Folding: "The process by which proteins acquire their functional, preordained, three-dimensional structure after they emerge, as linear polymers of amino acids, from the ribosome."

Proteomics: The term proteome refers to all the proteins expressed by a genome, and thus proteomics involves the identification of proteins in the body and the determination of their role in physiological and pathophysiological functions. ... Ultimately it is believed that through proteomics new disease markers and drug targets can be identified that will help design products to prevent, diagnose and treat disease.

Quantum: Describes a system of particles in terms of a wave function defined over the configuration of particles having distinct locations is implicit in the potential energy function that determines the wave function, the observable dynamics of the motion of such particles from point to point. In describing the energies, distributions and behaviours of electrons in nanometer-scale structures, quantum mechanical methods are necessary. Electron wave functions help determine the potential energy surface of a molecular system, which in turn is the basis for classical descriptions of molecular motion. Nanomechanical systems can almost always be described in terms of classical mechanics, with occasional quantum mechanical corrections applied within the framework of a classical model.

QuantumBrain: [theoretical] Think of your brain. Now, think of your brain performing at vastly superior levels. Nanobots will become an as-needed addition to your existing neurons, extending your mental capacities further then you can probably now imagine.

Quantum Computer: A computer that takes advantage of quantum mechanical properties such as superposition and entanglement resulting from nanoscale, molecular, atomic and subatomic components. Quantum computers may revolutionize the computer industry in the not too distant future.

Quantum Confined Atoms (QCA): atoms caged inside nanocrystals. May find uses in clear-glass sunglasses, bio-sensors, and optical computing.

Quantum Cryptography: A system based on quantum- mechanical principles. Eavesdroppers alter the quantum state of the system and so are detected. Developed by Brassard and Bennett, only small laboratory demonstrations have been made.

Quantum Dots: nanometer-sized semiconductor crystals, or electrostatically confined electrons. Something (usually a semiconductor island) capable of confining a single electron, or a few, and in which the electrons occupy discrete energy states just as they would in an atom (quantum dots have been called "artificial atoms"). [CMP] Other terminology reflects the preoccupations of different branches of research: microelectronics folks may refer to a "single-electron transistor" or "controlled potential barrier," whereas quantum physicists may speak of a "Coulomb island" or "zero-dimensional gas" and chemists may speak of a "colloidal nanoparticle" or "semiconductor nanocrystal." All of these terms are, at various times, used interchangeably with "quantum dot," and they refer more or less to the same thing: a trap that confines electrons in all three dimensions.

Quantum Dot Nanocrystals (QDNs): used to tag biological molecules, and "measuring between five and ten nanometres across, are made up of three components. Their cores contain paired clusters of atoms such as cadmium and selenium that combine to create a semiconductor. This releases light of a specific colour when stimulated by ultraviolet of a wide range of frequencies. These clusters are surrounded by a shell made of an inorganic substance, to protect them. The whole thing is then coated with an organic surface, to allow the attachment of proteins or DNA molecules. By varying the number of atoms in the core, QDNs can be made to emit light of different colours."

Quantum Mechanics: A largely computational physical theory explaining the behavior of quantum phenomena, which incorporates the theory of special relativity. Despite dilignet attempts, general relativity has not been sucessfully incorporated into quantum mechanics.

Quantum Mirage: A nanoscale property that may allow information to be transfered through use of the wave property of electrons. Thus, quantum computers might not require wires as we know them.

Quantum Tunneling: When electrons pass through a barrier, without overcoming it or breaking it down.

Quantum Well: A P-N-P junction in which the "N" layer is ~10 nm (where traditional physics leaves off and quantum effects take over) and an "electron trap" is created. "If one makes a heterostructure with sufficiently thin layers, quantum interference effects begin to appear prominently in the motion of the electrons. The simplest structure in which these may be observed is a quantum well, which simply consists of a thin layer of a narrower-gap semiconductor between thicker layers of a wider-gap material."

Quantum Wire: Another form of quantum dot, but unlike the single-dimension "dot," a quantum wire is confined only in two dimensions - that is it has "length," and allows the electrons to propagate in a "particle-like" fashion. Constructed typically on a semiconductor base, and (among other things) used to produce very intense laser beams, switchable up to multi-gigahertz per second.

Qubit: The quantum computing analog to a bit. Qubits exhibit superposition. Thus, unlike normal bits, qubits can be both 1 and 0 at the same time.

Red Goo: Deliberately designed and released destructive nanotechnology, as opposed to accidentally created grey goo.

Replicator: A system able to build copies of itself when provided with raw materials and energy.

Rheology: the study of the deformation and flow of matter under the influence of an applied stress, which might be, for example, a shear stress or extensional stress. The experimental characterisation of a material's rheological behaviour is known as rheometry, although the term rheology is frequently used synonymously with rheometry, particularly by experimentalists. Theoretical aspects of rheology are the relation of the flow/deformation behaviour of material and its internal structure (e.g. the orientation and elongation of polymer molecules), and the flow/deformation behaviour of materials that cannot be described by classical fluid mechanics or elasticity. This is also often called Non-Newtonian fluid mechanics in the case of fluids.

SAMFET: (self assembled monolayer field effect transistor). Where a few molecules act as FETs, exhibiting both very strong gain, and extraordinarily rapid response.

Scanning Capacitance Microscopy: A method for mapping the local capacitance of a surface.

Scanning Electron Microscopy: An instrument able to image surfaces to molecular accuracy by mechanically probing their surface contours. A kind of proximal probe. .... A device in which the deflection of a sharp stylus mounted on a soft spring is monitored as the stylus is moved across a surface. If the deflection is kept constant by moving the surface up and down by measured increments, the result (under favorable conditions) is an atomic-resolution topographic map of the surface. Also termed an atomic force microscope.

Scanning Near Field Optical Microscopy: A method for observing local optical properties of a surface that can be smaller than the wavelength of the light used.

Scanning Thermal Microscopy: A method for observing local temperatures and temperature gradients on a surface.

Scanning Tunneling Microscope (STM): An instrument able to image conducting surfaces to atomic accuracy; has been used to pin molecules to a surface.

Sealed Assembler Laboratory: A work space, containing assemblers, encapsulated in a way that allows information to flow in and out but does not allow the escape of assemblers or their products.

Self-assembly: In chemical solutions, self-assembly (also called Brownian assembly) results from the random motion of molecules and the affinity of their binding sites for one another. Also refers to the joining of complementary surfaces in nanomolecular interaction.

Self-repair: indicating ability to heal itself without outside intervention.

Self-replication: More accurately labeled "exponential replication," self-replication refers to the process of growth or replication involving doubling within a given period. Example: create one assembler. Program it to create another, and program that one likewise, etcetera, until you have a specified amount [which is the important part -- how to make them STOP].

Sentience Quotient: In the article "Xenopsychology" by Robert Freitas in Analog of April 1984 there is an interesting index called "Sentience quotient". It is based on: The sentience of an intelligence is roughly directly related to the amount of data it can process per unit time and inversely to the overall mass needed to do that processing. This would be something like baud/ kilograms. And since that would rapidly turn into a real big number, base 10 logs are used. The "least sentient" would be one bit over the lifetime of the universe massing the entire known universe, or about -70. The "most sentient" is claimed to be +50. Homo sapiens are around +13, a Cray I is +9, a venus flytrap is a peak of +1 with plants generally -2.

Single-walled carbon nanotubes (SWNT) See Nanotubes and buckyballs

Shape Memory Alloys: (SMA's) are a unique class of alloys which are able to "remember" their shape and are able to return to that shape even after being bent. The ability is known as the shape memory effect. ... This property has lead to many uses of SMA from orthodontics and coffee makers to methods of controlling aircraft and protecting buildings from earthquake damage. ... The first SMA to be discovered and the most commonly used is called Nitinol.

Singularity: Defined by Vernor Vinge as the "postulated point or short period in our future when our self-guided evolutionary development accelerates enormously (powered by nanotechnology, neuroscience, AI, and perhaps uploading) so that nothing beyond that time can reliably be conceived. ...a future time when societal, scientific and economic change is so fast we cannot even imagine what will happen from our present perspective, and when humanity will become posthumanity." Another definition is the singular time when technological development will be at its fastest. A grand evolutionary leap.

Sky Hook: A long, very strong, cable in orbit around a planet which rotates around its center of mass in such a way that when one end is closest to the ground, its relative velocity is almost zero. It would function as a kind of space elevator; shuttle craft would anchor to the end and then be lifted into orbit where they would be released. It is closely related to the idea of a beanstalk.

Smartdust: also "Smartdust Motes" "...tiny, bottle-cap-shaped micro-machines fitted with wireless communication devices - that measure light and temperature [among other things, such as environmental monitoring, health, security, distributed processing and tracking - ed]. When clustered together, they automatically create highly flexible, low-power networks with applications ranging from climate-control systems to entertainment devices that interact with handheld computers."

Smart Materials: Here, materials and products capable of relatively complex behavior due to the incorporation of nanocomputers and nanomachines.

Also used for products having some ability to respond to the environment. If you combined microscopic motors, gears, levers, bearing, plates, sensors, power and communication cables, etc., with powerful microscopic computers, you have the makings of a new class of materials: "smart materials." Programmable smart materials could shape-shift into just about any desired object. A house made of smart materials would be quite useful and interesting. Imagine a wall changing color at your command, or making a window where their was none before.

Space Fountain: A vertical stream of magnetically accelerated pellets reaching out into space, where a station held aloft by its momentum reverses the direction and directs it towards a receiver on the ground. Essentially a simpler version of a Lofstrom loop.

Spike : Another term for the singularity, suggested by Damien Broderick since the growth curves look almost like a spike as it is approached.

Spintronics: [AKA: Quantum Spintronics, Magnetoelectronics, Spin Electronics. Electronic devices that exploit the spin of electrons as well as their charge. Unlike conventional electronics which is based on number of charges and their energy, and whose performance limited in speed and dissipation, spintronics is based on the direction of electron spin, and spin coupling, and is capable of much higher speed at much lower power.

Star Trek scenario: Someone builds potentially dangerous self-replicating devices that spread disastrously.

Stewart Platforms: A positional device.

Superconductor: An object or substance that conducts electricity with zero resistance.

Superintelligence: An intellect that is much smarter than the best human brains in practically every field, including scientific creativity, general wisdom and social skills. This definition leaves open how the superintelligence is implemented: it could be a digital computer, an ensemble of networked computers, cultured cortical tissue or what have you. It also leaves open whether the superintelligence is conscious and has subjective experiences.

Superlattice Nanowire Pattern (transfer): [SNAP] a technique for producing "Ultra High Density Nanowire Lattices and Circuits".

Superposition: A quantum mechanical phenomena in which an object exists in more than one state simultaneously.

Superlattices: Artificial metallic superlattices are multilayered thin films, prepared by alternately depositing two elements using vacuum deposition or sputtering techniques. A wide spectrum of elements and compounds are suitable for deposition into superlattice structures, and the range of properties displayed by the resulting superstructures is greatly dependent upon the properties of both individual lattices as well as the interaction between them

Superlattice Nanowire : interwoven bundles of nanowires using substances with different compositions and properties.

Synthespian: An artificial actor, for example a 3D model animated by motion capture from a real actor or a computer program.

Technocyte: A nanoscale artificial device (especially a nanite) in the human bloodstream used for repairs, cancer protection, as an artificial immune system or for other uses.

Technofobics: Those who have a phobia to technology, and/or to advances in technology.

Terraform: To change the properties of a planet to make it more earthlike, making it possible for humans or other terrestrial organisms to live unaided on it, for example by changing atmospheric composition, pressure, temperature or the climate and introducing a self-sustaining ecosystem. This will most probably be a very long-term project, probably requiring self-replicating technology and megascale engineering. So far Venus and especially Mars looks as the most promising candidates for terraforming in the solar system. Speculation exists that with the advent of mature MNT that we should be able to accomplish Terraforming a planet such as Mars in years, rather then decades .

Thermal Noise: the vibration and motion of atoms and molecules caused by the fact that they have a temperature above absolute zero. [RCM] Once used as an argument on why MNT could not work

Top Down Molding: [AKA: mechanical nanotechnology] Carving and fabricating small materials and components by using larger objects such as our hands, tools and lasers, respectively. Opposite of Bottom up

Transhuman: someone actively preparing for becoming posthuman. Someone who is informed enough to see radical future possibilities and plans ahead for them, and who takes every current option for self-enhancement.

Transhumanism: Philosophies of life (such as Extropianism) that seek the continuation and acceleration of the evolution of intelligent life beyond its currently human form and human limitations by means of science and technology, guided by life-promoting values.

Transistor: the basic element in an integrated circuit. An on/off switch (consisting of three layers of a semiconductor material) that consists of a source (where electrons come from), a drain (where they go) and a gate that controls the flow of electrons through a channel that connects the source and the drain. There are two kinds of transistor, the bipolar transistor (also called the junction transistor), and the field effect transistor (FET).

Tribology: study of friction, wear and lubrication of interacting surfaces.

Tubeologist: Someone who knows their nanotubes inside and out,

Turing Test: Turing's proposed test for whether a machine is conscious (or intelligent, or aware): we communicate via text with it and with a hidden human. If we can't tell which of our partners in dialogue is the human, we say the computer is conscious.

Ubiquitous Computing: Also known as "embodied virtuality", "smart environment" and "ambient intelligence". Computers that are an integral, invisible part of people's lives. In some ways the opposite of virtual reality, in which the user is absorbed into the computational world. With ubiquitous computing, computers take into account the human world rather than requiring humans to enter into the computer's methods of working.

Universal Assembler: Uses raw atoms and molecules to construct consumer goods, and is pollution free. Can be programmed to build anything that is composed of atoms and consistent with the rules of chemical stability. Eric Drexler talks about these assemblers as nanorobots with telescoping manipulator arms that are capable of picking up individual atoms, and combining them however they are programmed.

Universal Constructor: A machine capable of constructing anything that can be constructed. The physical analog of a "universal computer", which can perform any computation.

Uplift: To increase the intelligence and help develop a culture of a previously non- or near-intelligent species.

Upload: (a) To transfer the consciousness and mental structure of a person from a biological matrix to an electronic or informational matrix (this assumes that the strong AI postulate holds). The term "Downloading" is also sometimes used, mainly to denote transferring the mind to a slower or less spacious matrix. (b) The resulting infomorph person. The origin of the term is uncertain, but obviously based on the computer technology term 'uploading' (loading data into a mainframe computer).

Utility Fog: [AKA: Polymorphic Smart Materials] Objects formed of "intelligent" polymorphic (able to change shape) substances, typically having an octet truss structure. Concept concieved by Dr. J. Storrs Hall.

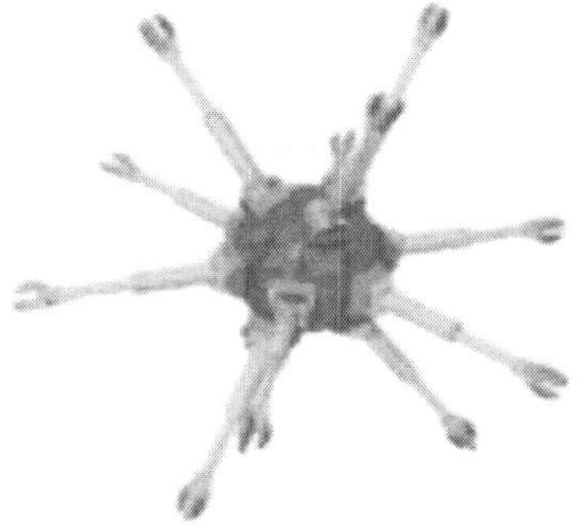

Single Foglet

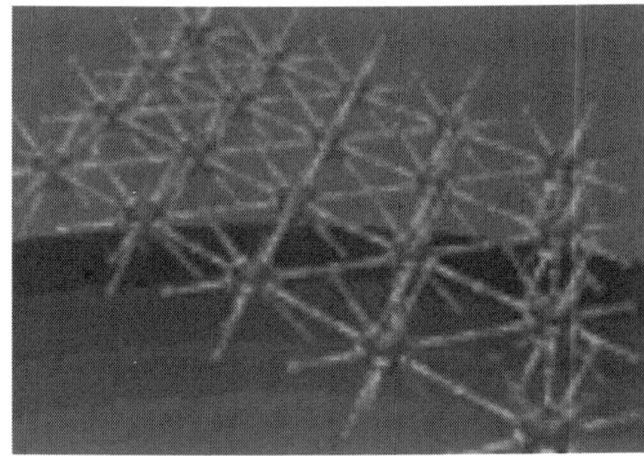

Foglets Holding Hands

"Imagine a microscopic robot. It has a body about the size of a human cell and 12 arms sticking out in all directions. A bucketfull of such robots might form a 'robot crystal' by linking their arms up into a lattice structure. Now take a room, with people, furniture, and other objects in it -- it's still mostly empty air. Fill the air completely full of robots. The robots are called Foglets and the substance they form is Utility Fog, which may have many useful medical applications. And when a number of utility foglets hold hands with their neighbors, they form a reconfigurable array of 'smart matter.'"

"Here's a short list of the powers you'd have or appear to have if embedded in fog: Creation--causing objects to appear and disappear on command. Levitation--causing objects to hover and fly around. Manipulation--causing forces (squeezing, hitting, pulling) on objects (real ones) at a distance. Teleportation--nearly any combination of telepresence and virtual reality between fog-filled locations."

Vasculoid: The vasculoid [concept] is a single, complex, multisegmented nanotechnological medical robotic system capable of duplicating all essential thermal and biochemical transport functions of the blood, including circulation of respiratory gases, glucose, hormones, cytokines, waste products, and cellular components

Virtual Nanomedicine: Using VR to perform surgery and other functions inside the body.

Von Neumann Machine: (pronounced von noi-man) A machine which is able to build a working copy of itself using materials in its environment. This is often proposed as a cheap way to mine or colonize the entire solar system or galaxy. An early fictional treatment was the short story "Autofac" by Philip K. Dick, published in 1955, which actually seems to precede John von Neumann's original paper about self-reproducing machines

VON Neumann Probe: A von Neumann Machine able to move over interstellar or interplanetary distances and to utilize local materials to build new copies of itself. Such probes could be used to set up new colonies, perform megascale engineering or explore the universe.

Wet Nanotechnology: the study of biological systems that exist primarily in a water environment. The functional nanometer-scale structures of interest here are genetic material, membranes, enzymes and other cellular components. The success of this nanotechnology is amply demonstrated by the existence of living organisms whose form, function, and evolution are governed by the interactions of nanometer-scale structures.

Zeptosecond: one-billion-trillionth of a second or 10 -21 second. Because nuclear movement takes place so quickly, scientists would need a pulse of light lasting just one zeptosecond to observe them.

Zettatechnology: in which zetta means 1021, referring to the typical number of distinct designed parts in a product made by the systems we envision (molecular, mature, or molecular-manufacturing-based nanotechnology). The term refers to the implemented technology and its products, rather than to intermediate steps on the pathway.

❑❑❑

References

Adleman, L.M. 1994. Molecular computation to solutions of combinatorial problems. Science, 266 : 1021 - 1024.

Ageev, B.G., Kapitanov, V., Ponomarev, Y.N., Vasilieve, V.A., Karapuzikov, A.I. and Scherstov, I.V. 2006. Laser gas-analyzers for studying kinetics of gas-exchange between vegetation biosystems and atmosphere. Proc. SPIE 6580 : 8.

Ahmed, I.A., Trau, M., Manne, S., Honma, I., Yao, N., Zhou, L., Fenter, P., Eisenberger, P.M. and Gruner, S.M. 1996. Biomimetic pathways for assembling inorganic thin films. Science, 273 : 892-899.

Amall, A.M. 2003. Future Technologies. Today's Choices : Nanotechnology, Artifical Intelligence and Robotics : A technical, political, and institutional map of emerging technologies, Greenpeace Environmental Trust, London.

Amato, I. 1999. Nanotechnology : Shaping the World Atom by Atom. National Science and Technology Council, Intergency Working Group on Nanoscience, Engineering and Technology, Washington, DC.

Arntz, Y., Sellig J.D., Lang, H.P., Zhang J., Hunziker, P. and Ramseyer, J.P. 2003. Label-free protein assay based on a nanomechanical cantilever array. Nanotechnology, 14 : 86-90.

Baeumner, A.J., Cohen, R.N., Miksic, V. and Min, J. 2003. RNA biosensor for the rapid detection of viable Escherichia coli in drinking water. Biosens Bioelectron, 18 : 405-413.

Bainbridge, W.S. and Roco, M.C. (Ed.) 2006. Managing Nano-bio-info-cogno innovations - converging technologies in society. Springer, The Netherlands, pg-390.

Balakrishna Rao, M. and Krishna Reddy. 2007. Nanotechnology in environment. Campus Books International, New Delhi, pg. 283.

Bandyopadhyay, A.K. 2010. Nanomaterials (II ed.). New Age International (P) Ltd., Publishers, New Delhi, pg. 308.

Bauer, G., Pittner, F. and Schalkhammer, Th. 1999. Metal nano-cluster biosensors. Mikrochim Acta, 131 : 107-114.

Balaji, S. 2010. Nanobiotechnology. MJP publishers. Chennai.

Birge, R.R. 1995. Protein based computers. Sci. Am. 90-95.

Boger, P. 1989. New plant-specific targets for future herbicides. In : Target Sites of Herbicide Action (ed. By Boger P. and Sandmann G.). CRC Press, Boca Raton, 247-282.

Booker. R. and Boysen, E. 2005. Nanotechnology for Dummies, Wiley Publishing.

Branda Paz. 2008. A handbook on nanoelctronics. Dominant Publisher and Distributors. New Delhi, pg 200.

Cai., H, Xu, C., He, P. and Fang, Y. 2001. Colloid Au-enhanced DNA immobilization for the electrochemical detection of sequence-specific DNA. J Electroanal Chem, 510 : 78-85.

Chakraborty, A. 2008. Nanotechnology - An introduction. Rajat Publications, New Delhi, pg. 297.

Chan, S., Fauchet. P.M., Li, Y., Rothberg, L.J. and Miller, B.L. 2000. Porous microcavities for biosensing applications. Phy Status Solid Appl Res., 182:541-546.

Chaudhary, M., Pandey, M.C., Radhakrishna, K. and Bawa, A.S. 2005. Nanotechnology : Application in food industry. Indian Food Industry, 24 : 19-21.

Chattopadhyay K.K. and Banerjee A.N. 2009. Introduction to nanoscience and nanotechnology. PHI learning Pvt., Ltd., New Delhi.

Chemla, Y.R., Grossman, H.L., Poon, Y., McDermott, R., Stevens , R. and Alpert, M.D. 2000. Ultrasensitive magnetic biosensors for homogeneous immunoassay. Proc Natl Acad Sci USA.m., 97: 14268 - 14272.

Choi,J.,Lim,. I.H., Kirn, H.H., Min. J. and Lee. W.H. 2001. Optical peroxide biosensor using the electrically controlled-release technique. Biosens Bioelectron.. 16 : 141 -146

Colvin, V.L.2003. 'The potential environmental impact of engineered nanomaterials'. Nature Biotechnology, 21 :1166-1170.

Cornell, B.A., Braach-Maksvytis, V.L.B., King L.G., Osman. P.O..1.,Raguse, B. and Wieczorek, L. 1997. A biosensor that uses ion-channel switches. Nature, 387: 580-583.

Cornell, B.A.. Krishna, G., Osman, P.O., Pace, R.D. and Wieczorek, L. 2001. Tethered bilayered lipid membrances as a support for membrane-active peptides. Biochem soc Trans.,29 : 613.

Cui, Y.,Wel. Q., Park, H. and Lieber, C.M. 2001. Nanowire nanosensors for highly sensitive and selective detection of biological and chemical species. Science, 293 : 1289 - 1292.

Cullum, B., Griffin, G., Miller, G. and Vo-Dinh. T. 2000. Intracellular measurements in mammary carcinoma cells using fiberoptic nanosensors. Anal Biochem., 277: 25-32.

Dagani, R. 2007. Individual Surface Atoms Identified. Chemical & Engineering News, page 13. Published by American Chemical Society.

David E. Reisner. (Ed.) 2009. Bionanotechnology - global prospects. CRC press, Boca Raton, FL, Pg. 349.

Davis, J.J., Coleman, K.S., Azamian, B.R. and Bagshaw. C.B. 2003. Green MLH. Chemical and biochemical sensing with modified single walled carbon manotubes. Chem Eur J., 15 : 315-321.

Diwan, P. (Ed.) 2009. Handbook of nanotecnology. Pentagon press, New Delhi, pg. 1136.

Diwan, P. and Bharadwaj, A (Ed.). 2005. The Nanoscope, Encyclopedia of nanoscience and nanotechnology, vol.V. Nanorobotics, Pentagon Press. New Delhi. pg.230.

Eibl, R.H. and Moy, V.T. 2005. Atomic force microscopy measurements of protein - ligand interactions on living cells. Methods Mol Biol., 305 : 439 - 450.

Elaissari, A. (Ed.) 2008. Colloidal Nanoparticles in Biotechnology, John Wiley & Sons Inc., New Jersey, pg. 358.

Elliott, D. and Zhang, W. 2001. Field assessment of nanoparticles of groundwater treatment. Environ. Sci. Technol., 35 : 4922-4926.

ETC Group: 2003. The Big Down: Atomtech - Technologies Converging at the Nano-scale, Winnipeg, Canada.

Garcia Martinex, J. (Ed.) 2009. Nanotechnology for the energy challenge. Wiley-VCH Verlag GmbH & Co. KaA, Weinheim, pg. 477.

Giessibl, F. 2003. Advances in Atomic Force Microscopy. Reviews of Modern Physics, 75 : 949-983.

Glazier R.. Venkatakrishnan, R., Gheorghiu, F., Walata, L., Nash, R. and Zhang, W. 2003. Nanotechnology takes root. Civil Engineering, 73 : 64-69.

Gogotsi, Y. (Ed.) 2006. Carbon nanomaterials. CRC press, Boca Raton, London, pg. 326.

Gonzalez-Garcia, M.B., Fernandez-Sanchez, C. and Costa-Garcia. 2000. A Colloidal gold as an electrochemical label of streptavidin-biotin interaction Biosens Bioelectron.. 15 : 315-321.

Greco, R.S., Pring. F.B. and Smith, R.L. (Ed.) 2005. Nanoscale Technology in Biological Systems. CRC Press. Boca Raton, Florida, pg. 485.

Gusev, A.I. and Rempel, A.A. 2008. Nanocrystalline materials. Viva Books Pvt. Ltd., New-Delhi, pg. 351.

Hari Singh Nalwa (Ed.) 2005. Handbook of Nanostructured biomaterials and their applications in Nanobiotechnology. Vol. 1 Biomaterials. Americian Scientific Publishers, California, USA, pg. 520.

Hari Singh Nalwa. (Ed.) 2005. Handbookof Nano structured Biomaterials and their applications in Nanobiotechnology Vol. 2. Applications in nanobiotechnology. American Scientific Publishers, California, USA, pg. 482.

Haruyama, T. 2003. Micro-and nanobiotechnology for biosensing cellular responses. Adv Drug Deliv Rev.. 55 : 393-401.

Heller K.H. and Atkinson B. (eds.) 2007. Agricultural nanotechnology dominant publishers and distributors, New Delhi.

Hornyak, G.L., Dutta, J., Tibbals. H.F. and Rao. A.K. 2008 Introduction to Nanoscience, CRC Press, Boca Raton. FL. pg. 815.

Hornyak, G.L., Moora, J. J, Tibbals. H.F. and Dutta, J. 2009. Fundamentals of Nanotechnology. CRC press, Boca Raton, FL. pg. 780.

ICAR, 1999. ICAR - Vision 2020. Indian Council of Agcirultural Research, New Delhi, India. Ministry of Agriculture 2000. National Agricultural Policy.

Jain K.K. 2003. Nanodiagnostics : Application of nanotechnology in molecular diagnostics. Expert Rev Mol Diagn., 3 : 153-161.

Jensen, R., Thursby, J.G. and Thurby. M.C. 2003. 'Disclosure and licensing of University inventions : "The best we can do with that we get to work with" International Journal of Industrial Organization, 21 : 1271-1300.

Jianrong, C., Yuging, Y., Nonyue, H., Xiaohu, W. and Sijias, L. 2004. Nanotechnology and Biosensors. Biotechnology Advances, 22 : 505-518.

Jinghua, G. 2004. Synchrotron radiation, soft X-ray spectroscopy and nano-materials. Journal of Nanotechnology 1 : 1-21-2, 193-225.

Joseph, T. and Morrison, M. 2006. Nanotechnology in Agriculture and Food. Institute of Nanotechnology, United Kingdom.

Joy, B. 2000. 'Why the Future Doesn't Need Us', Wired 8: 238-262.

Kapitanov, V.A. and Ponomarev, Y.N. 2006. Laser methanometer measurements of methene emission by plants in aerobic conditions. Atmos. Oceanic Opt., 651980:354-358.

Karkare, M. 2008. Nanotechnology fundamentals and applications. I.K. International Publishing House Pvt, Ltd., New Delhi, pg. 236.

Keepler, F., Hamilton, J.T.G., Braz. M. and Rockmann. T. 2006. Methane emission. from terrestrial plants under aerobic conditions. Nature, 439 : 187-191.

Khatkhar, B.S. 2006. Nanotechnology in food processing. Processed Food Industry, 9:31-34.

Kim, J, 2010. Current Research in nanotechnology. Apple Academic Press Inc., Oakvilled, ON L6L OA2, pg. 304.

Kirschvink, J.L., Koyayashi-Kirschvink, A. and Woodford. B.J. 1992. Magnetite biomineralization in the human brain. Proc. Natl. Acad. Sci., USA 89 : 7683- 7687.

Kohler, M. and Fritzche, W. (Ed.) 2007. Nanotechnology, An introduction to nanostructuring techniques. Wiley - VCH Verlag GmbH & Co. KGaA, Weinheim. pg. 321.

Kolusheva, S., Kafiri. R., Katz, M. and Jelinek R. 2001. Rapid colourmetric detection of antibody-epitope recognition at a biomimetic membrane interface. J Am Chem Soc., 1213:417-422.

Kostoff, R.N., Koytcheff, R.G. and Lau. C.G.Y. Global nanotechnology research review. Current Sci., 92 : 1492-1498.

Kostoff, R.N., Stump, J.A., Johnson, D, Murday, J.S., Lau C.G.Y. and Tolles, W.M. 2006. The structure and infrastructure of global nanotechnology literature. J. Nanopart. Res.. 8 : 301-321.

Kumar, A. and Whitesides, G.M. 1993. Features of gold having micrometer to centimeter dimensions can be formed through a combination of stamping within elastomeric stamp and an alkanethiol ink followed by chemical etching. App. Phys. Lett., 63-2002-2004.

Lapshin, R.V. and Obyedkov, O.V. 1993. Fast-acting piezoactuator and digital feedback loop for scanning tunneling microscopes. Review of Scientific instuments, 64 : 2883-2887.

Lapshin, R.V. 1995.Analytical model for the approximation of hysteresis loop and its application to the scanning tunneling microscope. Review of Scientific Instuments. 66 : 4718 - 4730.

Lapshin, R.V. 2004. Feature - oriented scanning methodology for probe microscopy and nanotechnology. Nanotechnology, 15 : 1135-1151.

Lec. J.W., and Foote. R.S. (Ed.) 2009. Micro and Nano technologies in method and protocols bioanalysis. Humana press, London, pg. 668.

Lien. H. 2000. Nanoscale bimetallic particles for dehalogentation of halogenated aliphatic compounds. Unpublished Dissertation, Lehigh University, Bethlehem, Pennsylvania.

Lien, H. and Zhang, W. 1999. Reactions of chlorinated methanes with nanoscale metal particles in aqueous solutions. J.Environ. Eng.,125 : 1042-1047.

Lien, H. and Zhang, W.2001.Complete dechlorination of chlorinated ethenes with nanoparticles. Colloids Surfaces 191: 97-105.

Lin, V.S., Motesharei, K., Dancil, K.P., Sailor M.J.and Ghadiri, M.R. 1997. A porous silicon-based optical interferometric biosensor. Science, 278 : 840-843.

Liu, X., Feng, Z., Zhang, S., Zhang, J., Xiao, Q. and Wang, Y. 2006. Preparation and testing of cementing nano-subnano composites for slow-or controlled release of fertilizers. Scientia Agriculture Sinica, 39 : 1598-1604.

Marc, D. and Sophie. D–C. 2003. Immobilisation of glucose oxidase within metallic nanotubes arrays for application to enzyme biosensors. Biosens Biolectron., 18 : 943-951.

Masciangioli,T. and Zhang, W. 2003. Environmental nanotechnology : Potential and pitfalls. Environ. Sci. Technol., 37 : 102A-108A.

Mathur, K.P. 2007. Nano-science and nanotechnology. Rajat publications. New Delhi, pg. 272.

Maxwell, D., Taylor, M.J. and Nie, S. 2002. Self-assembled nanoparticle probes for recognition and detection of biomolecules. J Am Chem soc., 124 : 9606-9612.

McKendry, R., Zhang. J., Arntz, Y., Strunz, T., Hegner, M. and Lang, H.P. 2002. Multiple label-free biodetection and quantitative DNA-binding assays on a nanomechanical cantilever array. Proc Natl Acad Sci USA., 99 : 9783-9788.

Miao, Y., Qi, M., Zhan, S., He, N., Wang, J. an Yuan, C. 1999. Construction of a glucose biosensor immobilized with glucose oxidase in the film of polypyrole nanotubules.Anal Lett, 32 : 1287-1299.

Moore, J.C., Jin, H.M., Kuchner, O. and Arnold, F.H. 1997. Strategies for the *in vitro* evolution of protein function. Enzyme evolution by random recombination of improved sequences. J. Mol Biol, 272 : 336-347.

Muir, T.W., Dawson, P.E. and Kent, S.B.H. 1997. Protein Synthesis by chemical ligation of protected perpetides in aqueous solution. Meth. Enzymol, 289-266-298.

National Research Council, 1999. Our Common Journey : a transition toward Sustainability, National Academy of Sciences, USA, 1999, 363 pp.

National Research Council. 1996. National Science Education Standards, National Academy Press, Washington, DC.

Nicewarner-Pene, S.R., Freeman. R.G., Reiss, B.D., He, L., Pena, D.J., Walton, I.D., Cromer, R.. Keating, C.D. and Natan, M.J. 2001. Submicrometer metallic barcodes, Science, 294 : 137-141.

Noji, H., Yasuda, R., Yoshida, M. and Knosita, K. Jr. 1997. Direct observation of the rotation of F-1-AT Pase.Nature. 386 : 299-302.

NRC (National Research Council) 1996. Biomolecular self-assembling materials: Scientific and technological frontiers. Washington, DC : National Academy Press. NRC. 1994. Hierarchical structures in biology as a guidefor new materials technology Washington, DC : National Academy Press.

Pak, S.C., Penrose, W. and Hesketh. P.J. 2001. An ultrathin platinum film sensor to measure biomolecular binding. Biosens Bioelectron., 16 : 371-378.

Pariera. N. 2007. Nanotechnology for environmental remediation. Pearl Books, New Delhi, pg. 248.

Park, S.J., Taton. T.A. and Mirkin, C.A. 2002. Array-based electrical detection of DNA with nanoparticle probes. Sciences, 295 : 1503-1506.

Peter, A., Singer, Fobio Salamanca - Buentello, Abdallah and Daar, S. 2005. Issues in Sciences and Technology online. Harnessing Nanotechnology to Improve Global Equity. http:www.issues. Org/21.4/singer.html

Pradeep, T. 2007. Nano - The Essentials - Understanding nanoscienceand nanotechnology. Tata MC draw- Hill Publishing company Ltd., New Delhi, pg. 444.

Qutubudding, S.. Wiencek, J.M., Nabi, A. and Boo, J.Y. 1994. Hemoglobin extraction using cosurfactant-free nonionic microemulsions. Sci. and Technology, 29 : 923-929.

Rathi. R. (Ed.) 2007. Coreconcept of nanotechnology with application spectrum. SBS publishers and distributors Pvt. Ltd.. New Delhi, pg. 344.

Ratner M. and Ratner D. 2006. Nanotechnology - A gentle introduction to the next big idea. Dorling Kindersley (India) Pvt. Ltd., Delhi.

Renugopalakrishanan, V. and Randolph V. Lewis (Ed.) 2006.Bionanotechnology - proteins to nanodevices. Springer, Dordrecht, The Netherlands, pg. 296.

Richardson, J., Hawkins, P.and Luxston, R. 2001. The use of coated paramagnetic particles as a physical label in a magnetoimmunoassay. Biosens Bioelectron.,16 : 989-993.

Robert E. Evenson, Carl E. Pray and Mark W. Rosegrant. 1999. Agricultural Research and Productivity Growth in India. Research Report 109. International Food Policy Research Institute, 2033 K Street, N.W.,Washington, D.C. 20006-1002 U.S.A. A.P. 103.

Rousseau, D.L. and Jelinski, L.W. 1991. Biophysics. In : Encyclopedia of Applied PhysicsVol. 2, New York. VCH Publishers.

Rugar, D., Zuger, O., Hoen, S., Yannoni, C.S., Veith, H.M. and Kendrick, R.D. 1994. Force detection of nuclear magnetic resonance. Science, 264 : 1560-1563.

Sarid, D. 1991. Scanning Force Microscopy, Oxford Series in Optical and Imaging Sciences, Oxford University Press, New York.

Shanmugam, S. 2010. Nanotechnology. MJP publishers, Chennai.

Schnur, J.M., Price,R. and Rudolph, A.S. 1994. Biologically engineered microstructures - Controlled release applications. J. Controlled Release., 28 : 3-13.

Scott, N. and Chen, H. 2003. Science and Enginnering for Agriculture and food systems. National Planning Workshop, Nov. 18-19, 2002 Washington DC pg. 1-61.

Semi, N. and Chen, H. 2003. Science and Engineering for Agricultural on food systems. National Planning Workshop, Nov. 18-19, 2002 Washington DC pg.1-61.

Semi, P. and Kulkarni, V.S. 2011. Environmental nanotechnology. Alpha publications, New Delhi, pg 312.

Seventh Biophysical Discusssion. 1995. Molecular Motors : Structure, mechanics and energy transduction. Biophys. J., 68 : Supplement S.

Shah, M.A. and Ahmad, T. 2010. Principles of nano-sciences and nanotechnology. Narosa Publishing House, New Delhi, pg. 220.

Shenton, W., Pum. D., Sleytr. U.B. and Mann, S. 1997. Synthesis of cadmium sulphide superiattices using self-assembled bacterial S-layers. Nature, 389: 585-587.

Siddharath Vaidhya, 2007. Nanotechnology : Research and perspective. Pearl Books, New Delhi, pg. 280.

Smith, G.B. and Granqvist. C.G. 2011. Green Nanotechnology. Solutions for sustainability and energy in the built environment. CRC press, Boca Raton, FL, pg. 447.

Song, J., Cheng. Q., Zhu, S. and Stevens, R.C. 2002. ''Smart'' materials for biosensing devices : cell-mimicking supramolecular assemblies and colorimetric detection of pathogenic agents. Biomed Microdevices., 4 : 213-221.

Sosnowski, R.G., Tu, E., Butler, W.F., Connell, J.P.O. and Heller, M.J. 1997. Rapid determination of single base mismatch mutations in DNA hybrids by direct electric field control. P. Natl. Acad. Sci. USA. 94 : 1119-1123.

Sotiropoulou, S., Gavalas, V., Vamvakaki, V. and Chaniotakis, N.A. 2003. Novel carbon materials in biosensor systems. Biosens Bioelectron., 18 : 211-215.

Su, M., Li. S. and Dravida. V.P. 2003. Microcantilever resonance-based DNA detection with nanoparticle probes. Apple Phys Lett., 82 : 3562-3564.

Su, X.L. and Li, Y. Quantum dot biolabelling coupled with immunomagnetic separation for detection of *Escherichia coli* 0157 : H7. 2004. Anal. Chem., 76 : 4806-4810.

Thursby, J.G. and Thurby, M.C. 2003, 'University Licening and the Bayh-Dole Act'', Science, 301 : 1052.

Timp, G. (Ed.) 2005. Nanotechnology. Springer-Verlag, New York. Inc.,pg. 696.

Toth, E. Pubanz, D., Vauthey, S., Helm L. and Merbach, A.E. 1996. The role of water exchange in attaining maximum relativities for dendrimeric MRI contrast agents. Chemistry - A European Journal, 2 : 1607-1615.

Trivedi, P.C. 2008. Nanobiotechnology. Pointer Publishers, Jaipur, pg. 186

Tseng, A.A. (Ed.) 2008. Nanofabrication fundamentals and applications. WorldScientific Publishing Co., Pvt. Ltd., Singapore, pg 574.

Tuan, V.D. 2002. Nanobiosensors : probing the sanctuary of individual living cells. J Cell Biochem Suppl., 39 : 154-161.

Vayssieres,L. (Ed.) 2009. On solar hydrogen and nanotechnology. John Wiley and sons (Asia) Ptd. Ltd., Singapore, pg. 680.

Vidyasankar, S., Ru, M. and Arnold. F.H. 1997. F.H. 1997. Molecularly imprinted ligant - exchange adsorbents for the chiral separation of underivatized amino acids. J. Chromatography. A, 775 : 51-63.

Wakabayashi K. and Boger. P. 2004. Phytotoxic sites of action for molecular design of modern herbicides (Part 1) : The photosynthetic electron transport system. Weed Biology and Management, 4 : 8 - 18.

Wakabayashi, K. 1989. Herbicidal targets in photosynthetic light reaction and molecular design of new herbicides. Plant Protection 43 : 575-584 (in Japanese).

Wakabayashi, K. 1993. Targets of herbicides : mechanism of action of herbicides. In : Pharmaceutical Research and Development (2nd Series): Management and Development of Agrochemicals, Vol. (ed. By Yajima H., Iwamura H., Ueno T. and Kamoshita K.). Hirokawa Publishing Tokyo, 393-420 (in Japanese).

Wakabayashi, K. and Sato, Y. 1992a. Herbicides affecting the structure and function of chloroplasts (1). Japan Pesitcide Information 60 : 25-31.

Warad, H.C. and Dutta, J. 2005. Nanotechnology for agriculture and food systems - a view. Microelectronics, School of Advanced Technologies, Asian Institute of Technology, POBox, 4, Klong Luang, Pathumathani 12120. Thialand 1-14.

Ward, M.D. and Ebersole, R.C. 1996. US patent 5501986.

West, P. 1994. Introduction to Atomic Force Microscopy : Theory, Practice and Applications. Cambridge Univesity Press, Cambridge.

Wilson, M., Smith K.K.G., Simmons, M. and Raguse, B. 2005. Nanotechnology-Basic science and Emerging technologies. Overseas Press India Pvt.India, New Delhi, pg. 271.

Winningham, M.J. and Sogah, D.Y. 1997. A modular approach to polymer architecture control via catenation of prefabricated biomolecular segments. : Polymers containing parallel beta-sheets templated by a Iphenoxathin-based reverse turn mimic. Macromolecules, 30 : 862-876.

Wohlstadter, J.N., Wilbur. J. L., Sigal, G.B., Biebuyck, H.A., Billadeau, M.A. and Dong. L. 2003. Carbon nanotube-based biosensor. Adv Mater., 15 : 1184-1187.

Wolfbeis. O. 2006. Composite Material for Simultaneous and Contactless Luminescent Sensing and Imaging of Oxygen and Carbon Dioxide. Advanced Materials, 18 : 1511-1516.

Xu, X., Liu, S. and Ju, H. 2003. A novel hydrogen peroxide sensor via the direct electrochemistry of horseradish peroxidase immobilized on collodal gold modified screen-printed electrode. Sensors, 3:350-360.

Yin, H., Wang, M.D.. Svoboda, K., Landick, R., Gelles, J. and Block. S.M. 1995. Transcription against an applied force. Science, 270: 1653-1657.

Yu. S.J.M., Conticello, V.P.., Zhang. G.H., Kayser, C., Fournier, M.J., Mason, T.L., and Tirrell. D.A. 1997. Smectic ordering in solutions and films of a rod-like polymer owing to monodispersity of chain length. Nature, 389 : 167-170.

Zhang. B., Zhang, Z.J., Wang, B., Yan, J.. Li J.J. and Cai, S.M. 2001. Preparation of gold nano-arrayed electrode on silicon substrate and its electrochemical properties. Acta Chim Sin., 56 : 1932-1936.

Zhang, W. 2003. Nanoscale iron particles for environmental remediation : An overview. J.Nanoparticle Res., 5 : 323-332.

Zhang, W., Wang, C. and Lien, H. 1998. Catalytic reduction of chlorinated hydrocarbons by bimetallic particles. Catal. Today, 40 : 387-395.

Zhang, Y. and Seeman, N.C. 1994. The construction of a DNA truncated octahedron. J. Am. Chem. Soc., 116 : 1661-1669.

Zhao, J. TDaly, J.P. Henkens, R.W., Stonehuerner, J. and Crumbliss, A.L. 1996. A xanthine oxidase / collodal gold enzyme electrode for amperometric biosensor applications. Biosens Biolectron., 11 : 493.

Zhao, Y-D Zhang, W-D, Chen, H, Luo, Q-M. and Li. S.F.Y. 2002. Direct electrochemistry for horseradish peroxidase at carbon nanotube powder microelectrode. Sens Actuators, B : 87 : 168-172.

Zimdahl, R.L. 1999. Fundamentals of Weed Science, Second Editiom, Academic Press, London, UK.

❑❑❑